Atlas of Mylonites- and related microstructures

Rudolph A. J. Trouw · Cees W. Passchier ·
Dirk J. Wiersma

Atlas of Mylonites- and related microstructures

Springer

Prof. Dr. Rudolph A.J. Trouw
Universidade Federal do Rio de Janeiro
Depto. Geologia
Ilha do Fundao
Rio de Janeiro-RJ
Cidade Universitaria
Brazil
trouw@igeo.ufrj.br

Prof. Dr. Cees W. Passchier
Universität Mainz
Institut für Geowissenschaften
Becherweg 21
55099 Mainz
Germany
cees.passchier@uni-mainz.de
cpasschi@uni-mainz.de

Dirk J. Wiersma
Oudwijk 12
3581 TJ Utrecht
The Netherlands
info@pantafos.com

Additional material to this book can be downloaded from http:// extras .springer .com .

ISBN 978-3-642-03607-1 ISBN 978-3-642-03608-8 (eBook)
DOI 10.1007/978-3-642-03608-8
Springer Heidelberg Dordrecht London New York

Library of Congress Control Number: 2009939942

Typesetting and book design by Elisabeth Sillmann, www.blaetterwaldDesign.de Landau, Germany

Cover design: Bauer, Thomas

Printed on acid-free paper

Springer is part of Springer Science+Business Media (www.springer.com)

Preface

Mylonites are high-strain rocks from ductile shear zones that are amongst the most impressive deformation structures in nature. Although they are an important source of geological information, many mylonites are not easy to interpret, even with the help of thin sections. In practice, one needs to build up some experience with mylonites before reliable interpretation is possible. Most textbooks can only show one or two photographs of a specific feature, presenting the best examples but unfortunately, these do not always represent geological reality. The best way to learn to interpret mylonites and other fault rocks is to see many examples, including imperfect ones. This atlas provides some of the necessary experience through a large number of thin section photomicrographs of mylonites and related microstructural features. The photomicrographs demonstrate what mylonites look like under the microscope and how they can be interpreted with the help of thin sections.

The main objective of this book is to help students and professional geologists to understand and classify mylonites and to answer the following questions.

- From what parent rock did this mylonite derive?
- What were the metamorphic circumstances during mylonitisation?
- What was the intensity of deformation?
- What was the sense of shear?

The photomicrographs shown in this atlas were taken from thin sections in a collection we compiled over the years from field work and excursions, with the help and participation of many geologists. The list of names of all persons involved would be too long, and we like to thank them collectively. Here we acknowledge those who contributed in a special way: Henk Zwart kindled our interest in microtectonics. Richard White kindly lend us a number of beautiful mylonite thin sections (all from Western Australia); Leda Fraga organized a field trip in Roraima, northern Brazil, on which we took many samples used for this atlas; John Bradshaw did the same in New Zealand and Luís Alberto Fernandes in Rio Grande do Sul, southern Brazil. Rodrigo Peternel, Camilo Trouw, Felipe Medeiros, Ron Vernon and Martyn Drury provided thin sections and/or photos that contributed significantly to this atlas. Other people that we like to acknowledge for their help, giving samples or taking us to the field, are: Guido Gosso, André Ribeiro, Brígida Castro de Machuca, Mariana Zuquim, Hugo Polo, Rodrigo Vinagre, Henrique Roig, Saskia ten Grotenhuis, Alexis Rosa Nummer, José Ramon Martinez Catalan and Antonia Baldez. José Altino helped with the drawing of several figures. Tarcisio Abreu elaborated high quality thin sections. Elisabeth Sillmann took care of the typesetting and layout.

Funding for our research was provided by CNPq and CAPES (Brazil), the Schürmann Foundation (the Netherlands), DFG and DAAD (Germany).

We hope that the examples shown in this atlas will help students and geologists working with mylonites to interpret these beautiful and important deformation features correctly.

R. A. J. Trouw
C. W. Passchier
D. J. Wiersma

Contents

Chapter 1 | Introduction

1 Introduction

It is not our intention to present a thorough theoretical treatment on mylonites and related rocks in this atlas. This can be found in several textbooks (e.g. Shelley 1993; Snoke et al. 1998; Blenkinsop 2000; Vernon 2004; Passchier & Trouw 2005) and in many articles referred to in these books. Especially the excellent photographic atlas by Snoke et al. (1998) gives an extensive overview of current ideas on nomenclature and processes related to fault rocks. Why then did we produce another atlas of mylonites? We felt the need for a simple, accessible treatment of the subject for students and all geologists with an interest in mylonites. Microstructural terminology follows essentially Passchier & Trouw (2005), to which this atlas provides an illustrated addition. Below we give a short overview of the most relevant aspects of mylonites and associated rocks.

1.1 Definition of the term "mylonite"

The concept of a mylonite is not easy to catch in a simple and precise definition (see e.g. Mawer 1986). Snoke & Tullis (1998) provide an excellent discussion on the historical evolution of the meaning of the word mylonite and the understanding of fault rocks by geologists. Below we give a simplified version and our personal view on this subject. Although mylonite is frequently used as a rock name, the term has in fact more of a structural than a lithological significance. The term was coined by Lapworth (1885) to designate fine-grained schistose rocks in faults such as the Moine Thrust in Scotland, formed mainly by a process of crushing, dragging out and grinding, hence the word mylonite, derived from the greek μυλων, a mill. Lapworth was, however, aware that secondary processes such as recrystallisation of quartz and mica also played a role during mylonite formation. Improvement of the knowledge how grain-size reduction takes place by recovery and recrystallisation (summarised, for example, in Passchier & Trouw, 2005) led to a significant modification of the concepts of faults, shear zones, mylonites and mylonite zones. Shear zones are now understood as narrow high-strain zones that accommodate lateral displacement of adjoining volumes of rock. The nature of deformation in shear zones varies with depth, thermal gradient, rock type and several other factors (e.g. Ramsay 1980). Shear zones in which deformation was exclusively brittle are known as faults. Other shear zone segments can have a component of ductile deformation whereas many are exclusively ductile in their behaviour (ductile shear zones). For historical reasons the term fault is still used sensu lato in these cases to include narrow brittle-ductile and sometimes even ductile shear zones. With new insights on deformation mechanisms, crushed rocks from brittle faults are now called cataclasites or fault breccias and the term mylonite is restricted to cohesive rocks from ductile shear zones, generally formed under metamorphic circumstances.

Based on these considerations, a simple definition of a mylonite would be: "a strongly deformed rock from a ductile shear zone". Although this definition is essentially correct it is not without its problems: not all rocks from ductile shear zones are mylonites and some mylonites are difficult to associate with a specific ductile shear zone.

Ductile shear zones are recognized because strongly deformed rocks are bordered by less deformed rocks. Hence, why are not all rocks from ductile shear zones mylonites? One should realise that mylonites are metamorphic rocks and as such show mineralogy and structure related to their peak metamorphic conditions. If a mylonite was formed by strong ductile deformation before local peak metamorphic conditions, recrystallisation and grain growth might have completely overprinted the typical deformation structures of the mylonite, and it would no longer be recognisable as such. Therefore, if deformation in a shear zone generates mylonites and the temperature increases afterwards, the mylonitic aspect of the structure may be destroyed, but the shear zone may still be recognizable by dislocated markers.

A second problem is that some mylonites cannot easily be associated with specific ductile shear zones. This is often a matter of scale. If an entire mountain range is intensely deformed and deformation is not clearly constrained to distinct shear zones, all involved rocks are called mylonites because they share the same microstructure? After all one could argue that the entire mountain range could be considered a huge shear zone. However, this is not common practice.

We may conclude that the definition of mylonites as rocks from shear zones is not fully satisfactory (compare Wise et al. 1984; Mawer 1986; Snoke & Tullis 1998).

Another possible definition could simply be: "a mylonite is a strongly deformed rock". By definition the deformation in shear zones is stronger than in adjacent areas, hence the mylonites would be the strongest deformed rocks in such a context. Although this definition is attractive and often correct, it suffers from the very subjective significance of the word "strongly". What one scholar may consider a "strongly deformed rock" another may interpret as just "deformed"; and how can we judge the intensity of total deformation anyway from structures that usually only show the last increments of deformation?

As explained above, most metamorphic rocks undergo a complex evolution of repeated deformation, recrystallisation and grain growth. Metamorphic rocks usually seem to keep pace in this process and eventually form regular schists or gneisses. Some metamorphic rocks, however, suffer relatively intense and/or rapid deformation during the retrograde leg of their P-T-t paths and these would show up as mylonites. However, the quantity of total deformation in the schists may be as high as in these mylonites. This is sometimes obvious by the dislocation of schists in nappe complexes. Therefore, it is the intensity of deformation visible in the microstructure that classifies a rock as a mylonite, rather than the total strain that the rock has suffered.

We may conclude that the term mylonite refers to rocks with a specific (micro)structure that, in most cases, can be qualified as follows:

- Presence of a strong SL fabric
- Presence of a fine-grained matrix with porphyroclasts. Minerals like quartz, chlorite, biotite and muscovite are usually present in the matrix, either highly strained at low grade, or recrystallised at higher grades. Minerals like feldspar, garnet, hornblende and pyroxenes may form porphyroclasts, commonly showing evidence of crystal-plastic deformation by undulose extinction and/or partial recrystallisation
- Presence of a certain asymmetry, especially in low-grade mylonites, in the form of C/S fabric or C′ shear bands, mineral fish, stair stepping, oblique foliation etc.

Finally, we may define mylonites as rocks of any composition, usually associated with shear zones, with a specific structure indicative of stronger ductile deformation then adjacent rocks. This structure can be further specified by the presence of a strong SL fabric, the presence of a relatively fine-grained matrix with porphyroclasts (only absent in ultramylonites) and the frequent occurrence, especially in low-grade mylonites, of asymmetric structures like C/S or C′ shear bands, mineral fish, stair stepping, oblique foliation etc. To be preserved they tend to form on the retrograde leg of the P-T-t path.

1.2 Classification of mylonites

Mylonites have been subdivided into protomylonites, mylonites and ultramylonites, depending on the percentage of matrix versus porphyroclasts (Sibson 1977; Scholz 1990; Schmid & Handy 1991). A protomylonite has between 10 and 50% matrix, a mylonite has 50-90% matrix and an ultramylonite has between 90 and 100% matrix. Some ultramylonites have also been referred to as phyllonites, a term also used for mylonites rich in mica, derived from schists.

Since the mylonite matrix is usually recrystallised and porphyroclasts are normally not, the classification is sometimes based on the percentage of recrystallisation. However, this practice is not recommended because in low-grade mylonites a considerable part of the matrix is not necessarily the product of recrystallisation and in high-grade mylonites a considerable part of the porphyroclasts may be composed of recrystallised grains.

Although the classification of Sibson (1977) has proven to be useful, in practice the following remarks have to be considered. First, the precise limit in terms of grain size between matrix and porphyroclasts is subjective; this introduces a certain lack of precision into the classification. Secondly, the percentage of porphyroclasts is strongly dependent on the type of parent rock. For instance, a quartzite subject to mylonitisation will form an ultramylonite at much lower strain than a granite, because the first has no feldspar to form porphyroclasts. Hence, comparison between the three types of mylonites in terms of strain intensity is only relevant if the same parent rock is considered.

Another term often used in the literature on mylonites is blastomylonite, to designate those mylonites with static recrystallisation of part of the fabric after deformation ceased. Since most mylonites show at least some recrystallisation and since it is hard to define what part of this recrystallisation was static, we have chosen, in this atlas, to subdivide mylonite into low-grade, medium-grade and high-grade ones, mainly based on the grain size , type and completeness of recrystallisation.

1.3 Determining factors in the formation of mylonites

The formation of mylonites and related rocks is mainly controlled by the following variables:

- Intensity (and/or rate) of deformation
- Metamorphic conditions during and after mylonitisation; this includes temperature, confining pressure and the presence and composition of fluids
- Mineralogical composition of the parent rock

The first of these variables is expressed in the classification of protomylonites, mylonites and ultramylonites, of which examples are shown in Chapter 7. This classification is particularly useful in low-grade mylonites where clear contrasts between these three types are relatively common. In higher-grade mylonites with larger grain size in the matrix and more equal strain distribution, the transition between the three types is usually more gradual.

The second variable, metamorphic conditions, can also be understood as depth of formation (Fig. 1.1). Chapters 2 to 6 show examples of shear zone related rocks formed under conditions of increasing temperature and pressure. Cataclasites and fault breccias are typical of low temperatures, close to the surface (Chapter 2), where deformation occurs in the brittle regime. Pseudotachylytes (Chapter 3) form also in the brittle regime where earthquake-generating motion

Fig. 1.1 Distribution of the main types of fault rocks with depth in the crust.

a Schematic cross-section through a transcurrent shear zone. The zone may widen, and changes in geometry and dominant type of fault rock occur with increasing depth and metamorphic grade.

b Schematic representation of four typical fault rocks (out of scale) and the local geometry of the shear zone in a one-meter-wide block, such as would develop from a phenocryst granite. Inclined (normal or reverse) shear zones show a similar distribution of fault rocks and shear zone geometry with depth. No vertical scale is given since the depth of the transition between dominant ductile deformation and brittle fracturing depends on rock composition, geothermal gradient, bulk strain rate and other factors.
gs: generation surface.
(after Passchier & Trouw 2005, their Fig. 5.2)

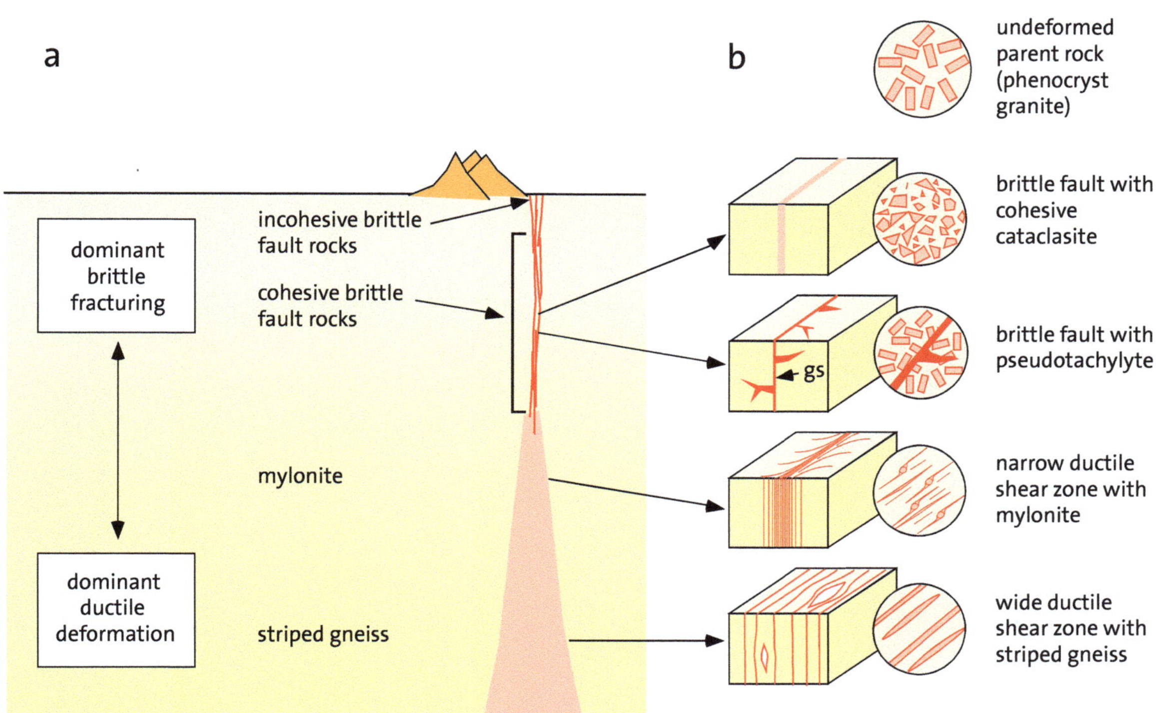

produces sufficient heat by friction to locally melt the wall rock or to melt cataclasite derived from the wall rock. Local melting produces veins of liquid that solidify quickly after the movement stops. Chapters 4, 5 and 6 show examples of mylonites, derived from quartzo-feldspathic rocks, that are formed under increasing temperature. Medium-grade mylonites often show a granoblastic texture of quartz in the matrix, with small to medium grain size (blastomylonites). High-grade mylonites, also referred to as striped gneisses, are usually characterized by the presence of few porphyroclasts. Strain free quartz grown to medium grain size often forms ribbons between recrystallised feldspar.

In some reactivated shear zones, lower grade mylonites can be superposed on higher grade ones showing structures of contrasting metamorphic grade. The inverse may also occur but will be rarely preserved.

Examples of the influence of the third variable, mineralogical composition, are shown in Chapter 8. Most examples are from mafic parent rocks, such as amphibolites, but examples of mylonitised peridotites, serpentinites, eclogites, calcsilicate rocks and carbonatic rocks are also shown.

Chapter 9 is dedicated to shear sense indicators, which are usually much better developed in low-grade mylonites than in high-grade ones.

Correct interpretation of the microstructure of quartz in the progressive process of crystal-plastic deformation, recovery and recrystallisation, is crucial to the understanding of mylonites. Therefore, a short review of these structures is given in Chapter 10. It has to be kept in mind, however, that these structures are not exclusive for mylonites and are therefore usually treated separately in textbooks.

Some rocks have structures that resemble mylonites closely, but that are formed under different circumstances. Examples of these are shown in Chapter 11 under the heading "false" mylonites.

Finally, a number of exercises are given in Chapter 12 to give the reader the opportunity to check his or her ability to interpret mylonites. We offer our interpretation at the end of the chapter, but are aware that in some cases other interpretations may be possible.

In addition to the cited references, an extensive bibliography lists most articles and books on the subject of mylonites and related rocks that were available at the end of 2008.

Samples of mylonites and other fault rock structures shown in photomicrographs in this book have been taken from shear zones in many different tectonic settings, in order to expose the reader to a large number of examples with different geometries, rock types and metamorphic conditions. We have purposely mixed dextral and sinistral shear sense in our examples, in order to make the approach more realistic.

Most thin sections have the location of origin mentioned in the caption; some are from unknown origin and others are from the "Leiden collection", the extensive teaching collection of thin sections of Henk Zwart at Leiden University, for which the location was not always recorded.

In general we have avoided using abbreviations but in the figure captions we use PPL for plain polarized light and CPL for crossed polarized light.

Chapter 2 | Cataclasites

2 Cataclasites

The cataclasites presented in this chapter are all cohesive fault rocks that show evidence of brittle fracturing although other processes such as grain boundary sliding and pressure solution may have played a role in their formation as well. These rocks are usually composed of angular broken rock fragments embedded in a matrix of quartz, iron oxide, calcite, chlorite and / or other minerals that precipitated from a fluid. Cohesive cataclasites are thought to form in the P-T realm where brittle deformation predominates, that is, approximately in the upper 10 km of the Earths crust, with lithostatic pressure up to about 3 kbar and temperatures up to about 300 °C (Fig. 1.1). However, this depth cannot be established with precision because other factors, like the presence or absence of fluids and the strain rate, play also an important role.

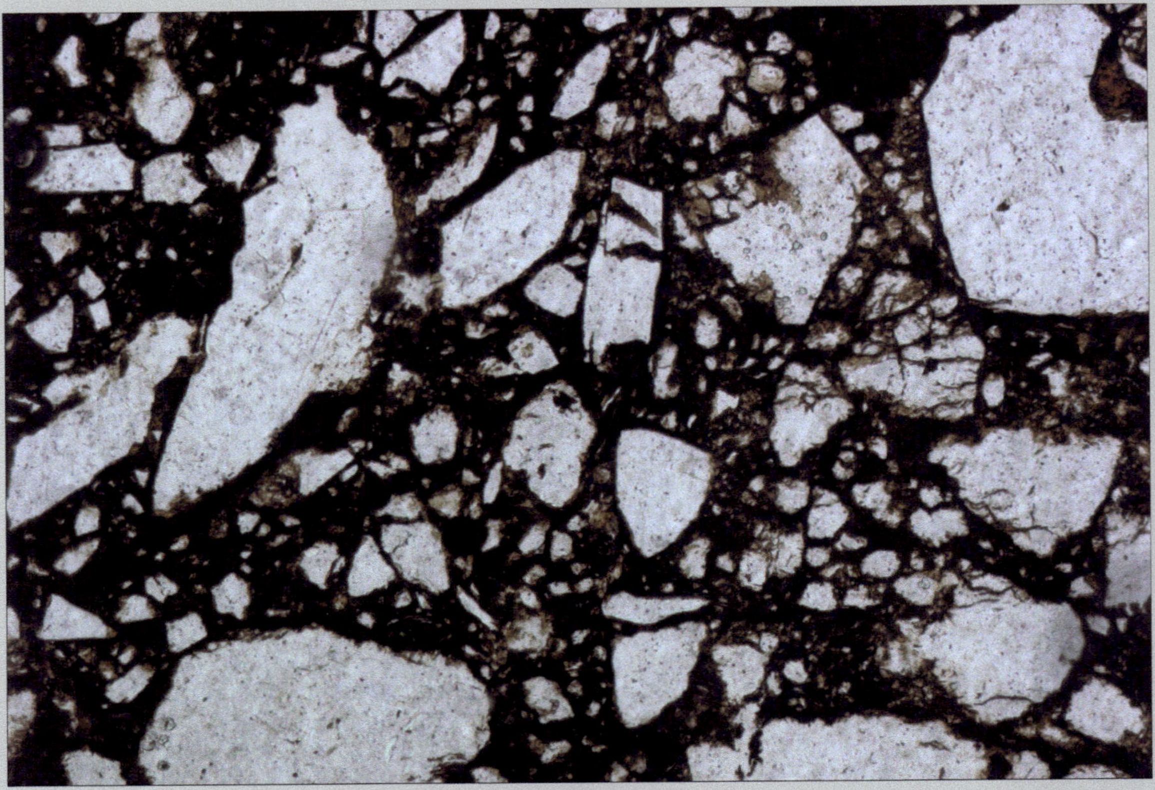

Fig. 2.1 Cataclasite showing angular rock fragments, mainly of quartz, embedded in a matrix of iron oxide. It is difficult to define a parent rock in this case, because the sample was taken from a fault zone that separates Precambrian gneisses from Tertiary sedimentary conglomerates and sandstones. Many of the quartz fragments are probably derived from quartz veins in the fault zone. Buzios, Rio de Janeiro State, SE Brazil.
Width of view 4 mm. Plain Polarized Light (PPL).

Fig. 2.2 Transition between a fractured rock at lower left to a fault breccia in the center and to a mica bearing cataclasite in the upper right hand part of the photograph. Notice the curved mica at right. Buzios, Rio de Janeiro-State, SE Brazil. Width of view 3 mm. PPL.

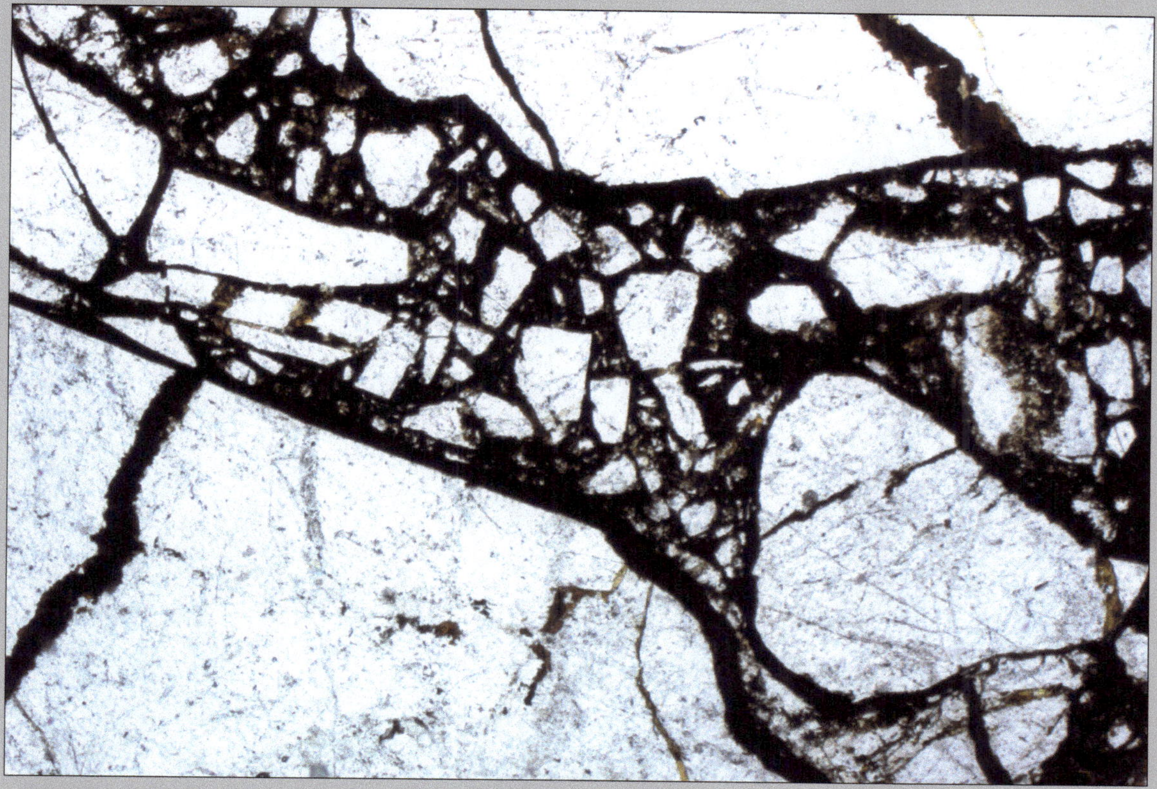

Fig. 2.3 Narrow zone of fault breccia and cataclasite in a quartzitic parent rock. The matrix is mainly composed of iron oxide. Note the lack of undulose extinction in the quartz fragments (Fig. 2.4). Buzios, Rio de Janeiro State, SE Brazil. Width of view 4 mm. PPL.

Fig. 2.4 As Fig.2.3. Crossed Polarized Light (CPL).

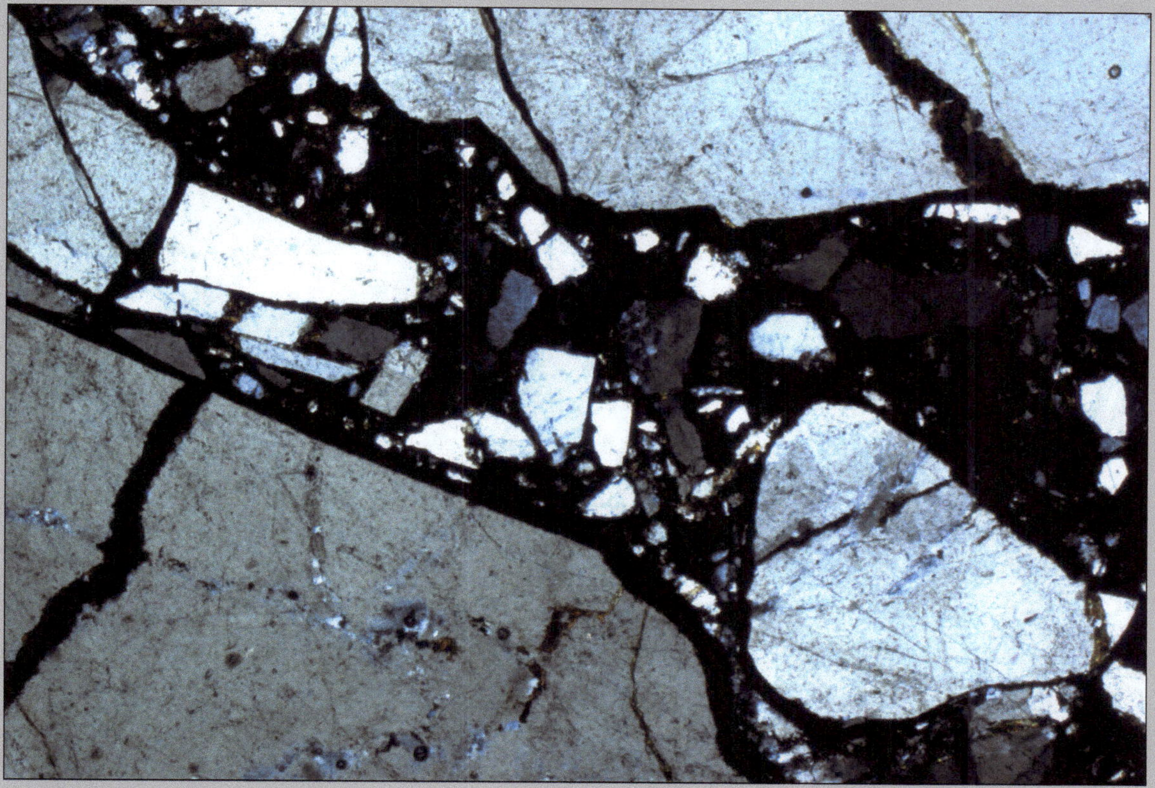

Fig. 2.5 Deformed quartzite with a cataclastic shear zone. Undulose extinction in the wall rock in the upper and lower parts of the photomicrograph shows that this rock experienced some ductile deformation. The horizontal brittle shear zone in the central part has relatively sharp contacts and is filled with fragments of quartz ranging from very small to medium grain size. In this case it is difficult to define the percentage of matrix of this cataclasite because the transition between fragments and matrix is gradational. No sense of shear can be determined from this structure. Guapé, southern Minas Gerais State, SE Brazil. Width of view 4 mm. CPL.

Fig. 2.6 Quartzite showing evidence of weak ductile deformation transected by a cataclastic shear zone. Note the large variation of grain size of the angular fragments in the shear zone. The quartz grains in the wall rock have interlobate contacts due to grain boundary migration and incipient bulging recrystallisation. Guapé, southern Minas Gerais State, SE Brazil. Width of view 4 mm. CPL.

Fig. 2.7 Deformed low-grade quartzite with a cataclastic shear zone in vertical position. Deformation in the quartzite is evident by the sweeping undulose extinction in most grains and by the lobate contacts due to low temperature grain boundary migration. The small grain size in the cataclastic shear zone is probably mainly the result of fracturing, but no clear matrix is developed. Other possible mechanisms that may have contributed are grain boundary sliding and dissolution. Guapé, southern Minas Gerais State, SE Brazil. Width of view 4 mm. CPL.

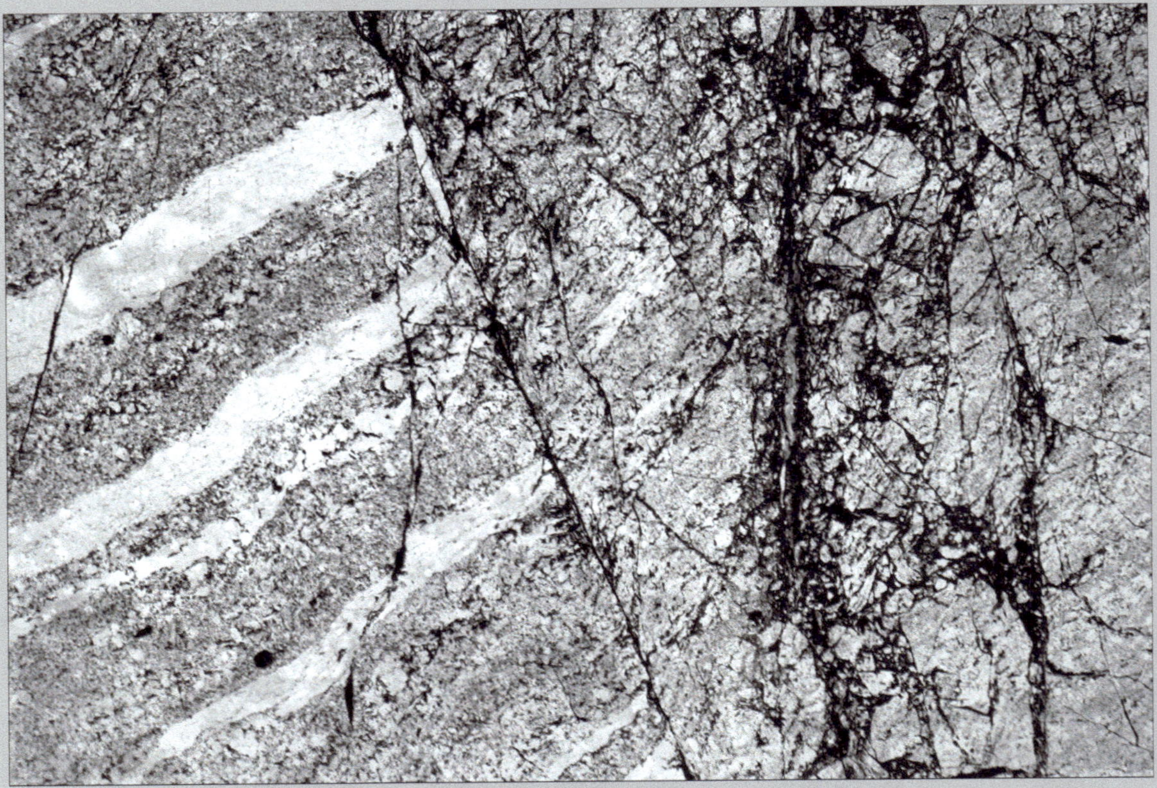

Fig. 2.8 Fault breccia, formed in a low-grade mylonite derived from a quartz-feldspathic rock. Notice the low percentage of matrix, defined as opaque material along the fault planes. No reliable sense of shear can be determined in this photomicrograph. Guapé, southern Minas Gerais State, SE Brazil. Width of view 22 mm. PPL.

Fig. 2.9 As Fig. 2.8. Width of view 22 mm. CPL.

Fig. 2.10 Cataclasite derived from a protomylonitic granite showing how the individual fragments are rotated during deformation, visible by the different orientation of their foliation. Pyrenees, Spain.
Width of view 20 mm. PPL.

Fig. 2.11 As Fig.2.10. Width of view 20 mm. CPL.

Fig. 2.12 Mylonitic quartz muscovite schist with horizontal schistosity, cut by a subvertical cataclastic vein, filled with rock fragments and abundant iron oxide. Some mica fish in the wall rock, to the right from the centre, indicate dextral sense of shear in the schist. No sense of shear can be determined in the cataclastic vein. Carrancas, southern Minas Gerais State, SE Brazil. Width of view 20 mm. CPL.

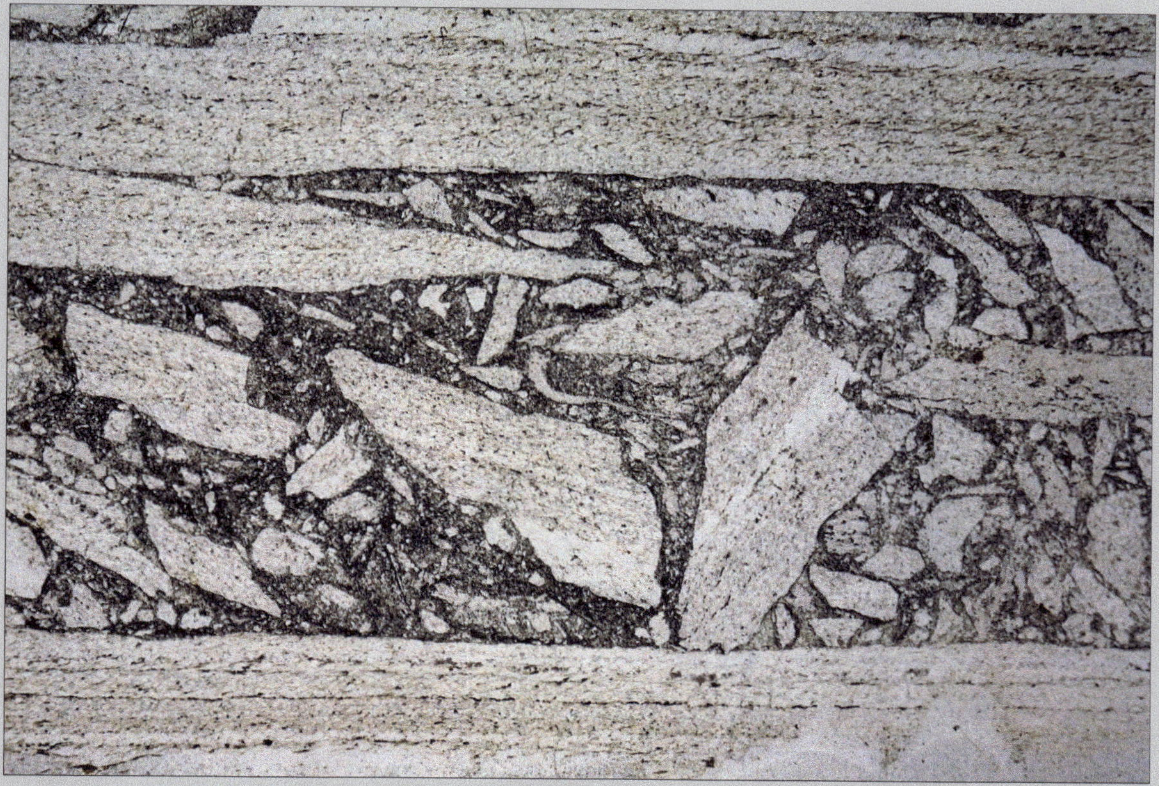

Fig. 2.13 Cataclasite developed in a mylonitic granite. The inclined orientation of elongated fragments at upper right probably indicates sinistral sense of shear. Uruguay. Width of view 16 mm. PPL.

Fig. 2.14 As Fig. 2.13. Width of view 16 mm. CPL.

Fig. 2.15 Low-grade ultramylonite, transformed into a fault breccia in the upper part of the photomicrograph. No reliable sense of shear can be determined from this image. The fabric may be due to superposition of low temperature brittle deformation on a structure formed at higher temperature. However, it is also possible that ductile deformation took place at low strain rate under low grade conditions and that the acceleration of this rate along a certain layer caused the fracturing at the same metamorphic circumstances. Uruguay. Width of view 20 mm. PPL.

Fig. 2.16 Dextral fault zone in a fine-grained sedimentary rock. Notice that no cataclasite or breccia is visible at this magnification. Mechanisms to make this kind of movement possible include grain boundary sliding and dissolution. Crato, NE Brazil. Width of view 20 mm. PPL.

Fig. 2.17 Fault breccia developing in a quartz-rich gneiss. The white bands are large quartz crystals. Undulose extinction of the grey quartz above the center is evidence of minor ductile deformation. San Juan, Argentina. Width of view 12 mm. CPL.

Fig. 2.18 Mylonite derived from an amphibolitic and biotite-rich parent rock, cut by a fault (upper left to lower right). Note the hornblende porphyroclast at lower left and the garnet porphyroclast in the center, cut in half by the fault. The fault plane is marked as a dark line of opaque material. No sense of shear is apparent. Scotland. Width of view 5 mm. PPL.

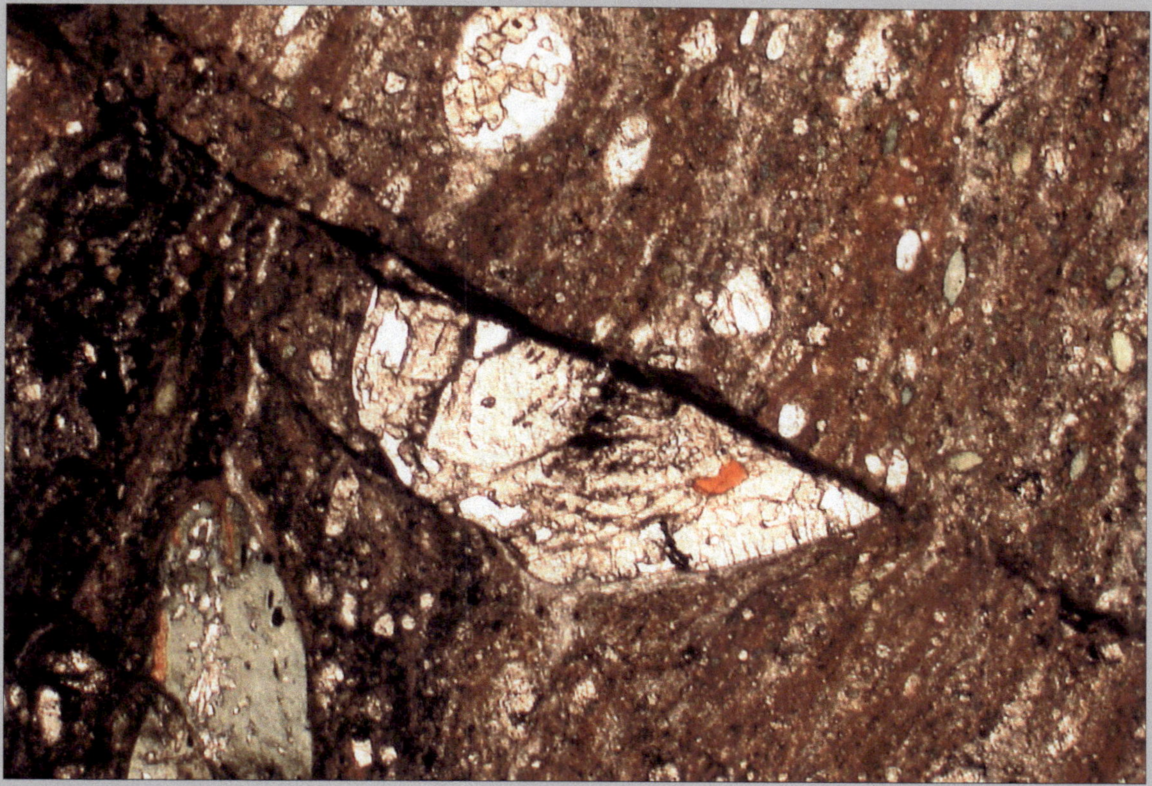

Fig. 2.19 Cataclastic zone in deformed granite. Note variation in size of fragments and the presence of sharp black lines representing fault planes (upper right to lower left). No sense of shear can be deduced from this view. Kaokoveld, Namibia. Width of view 16 mm. PPL.

Fig. 2.20 As Fig. 2.19. Width of view 16 mm. CPL.

Fig. 2.21 Amphibole-bearing granodiorite (upper part) cut by a cataclastic fault zone (lower part). Note the presence of angular rock fragments. A probable sinistral sense of shear can be inferred from the inclined lenses in the cataclasite close to the contact. San Juan, Argentina. Width of view 23 mm. PPL.

Fig. 2.22 As Fig. 2.21. Width of view 23 mm. CPL.

Fig. 2.23 Cataclastic to protomylonitic granite cut by a zone of cataclasite (subhorizontal through the center). The foliation in the lower part, inclined to the right, probably indicates sinistral sense of shear. Leiden collection. Width of view 15 mm. PPL.

Fig. 2.24 As Fig. 2.23. Width of view 15 mm. CPL.

Fig. 2.25 Cataclasite developed in an ultramylonite. The laminated rock visible in the upper part is an ultramylonite. Angular fragments are recognizable in the central part, embedded in a fine-grained greenish matrix composed of chlorite and quartz. The sample comes from a ductile shear zone that was later reactivated at lower temperature, in the brittle regime. Virginia, southern Minas Gerais State, SE Brazil. Width of view 20 mm. PPL.

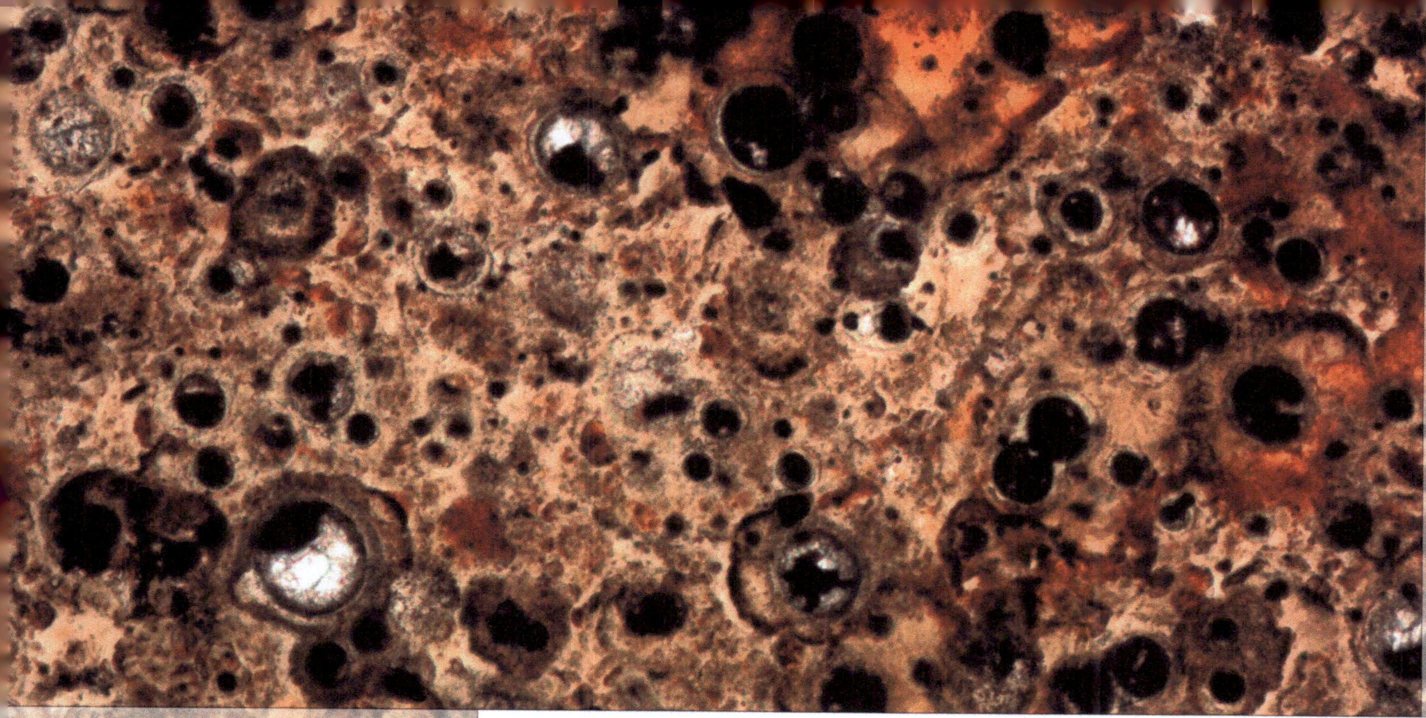

Chapter 3 | Pseudotachylytes

3 Pseudotachylytes

Pseudotachylytes are cohesive glassy or very fine-grained fault rocks that characteristically occur in veins. They are composed of an extremely fine-grained or glassy matrix with inclusions of wall-rock fragments. Commonly, a straight main fault vein (Sibson 1975; Spray 1992) representing a generation surface (Fig. 1.1) is present, from which smaller injection veins branch out. The main criteria to distinguish pseudotachylytes from cataclasites are:

- The presence of injection veins branching from a straight main fault vein (generation surface).
- A sharp transition between pseudotachylyte veins and the wall rock, with characteristic embayments at the site of mica or hornblende crystals; in cataclasites the transitions are more commonly gradual. There is potential confusion, however, where pseudotachylyte is in contact with cataclasite rather than with intact wall rock.

Other typical pseudotachylyte microstructures are a preferential occurrence of clasts of quartz and absence of mica clasts in the matrix, fine powdery sulphite in quartz clasts, and the presence of larger clasts in a homogenous fine-grained matrix. In cataclasite, there tends to be a more general transiton between matrix, small clasts and larger clasts.

Pseudotachylytes are thought to form by generation and subsequent melting of fine-grained cataclastic rock along friction surfaces. Shear stresses build up to high levels until they surpass the shear resistance of the rock, and stored elastic energy is then suddenly released by rapid motion on the fault (a process often referred to as "stick-slip") and released as heat and seismic waves, producing an earthquake. The generated heat melts the rock and cataclasite along the friction surfaces in a matter of seconds. However, this heat is restricted to a very narrow zone and is therefore quickly dissipated after the movement stops, causing chilling and solidification of the liquid to a glass or aphanitic crystalline material.

Pseudotachylytes form at the same depth and temperature range as cataclasite. At deeper levels of the crust higher temperature favours ductile deformation by crystal-plastic processes, generating mylonites along shear zones (Fig.1.1). Here the deformation tends to be more continuous with time.

Pseudotachylytes can also be associated with meteorite impact structures, and all pseudotachylyte veins of more than two centimeters thick should be investigated with this possibility in mind. In some cases pseudotachylytes suffer ductile deformation after their solidification. This causes deformation of the inclusions to produce a preferred shape orientation. Some examples of this process are shown below.

Fig. 3.1 Pseudotachylyte vein cutting a protomylonitic hornblende bearing granodiorite. Note the sharp contacts, the presence of inclusions and the black glassy matrix. No sense of shear can be established form this structure. Roraima, northern Brazil. Width of view 17 mm. PPL.

Fig. 3.2 As Fig. 3.1. Width of view 17 mm. CPL.

Fig. 3.3 Pseudotachylyte band in granite. Note the sharp contacts and the feldspar inclusions in the dark glassy matrix. No sense of shear is apparent. Roraima, northern Brazil. Width of view 5 mm. CPL.

Fig. 3.4 Contact of pseudotachylyte (upper part) and granitic country rock (lower part). Note the sharp contact with small injection veins in the center. Roraima, northern Brazil. Width of view 5 mm PPL.

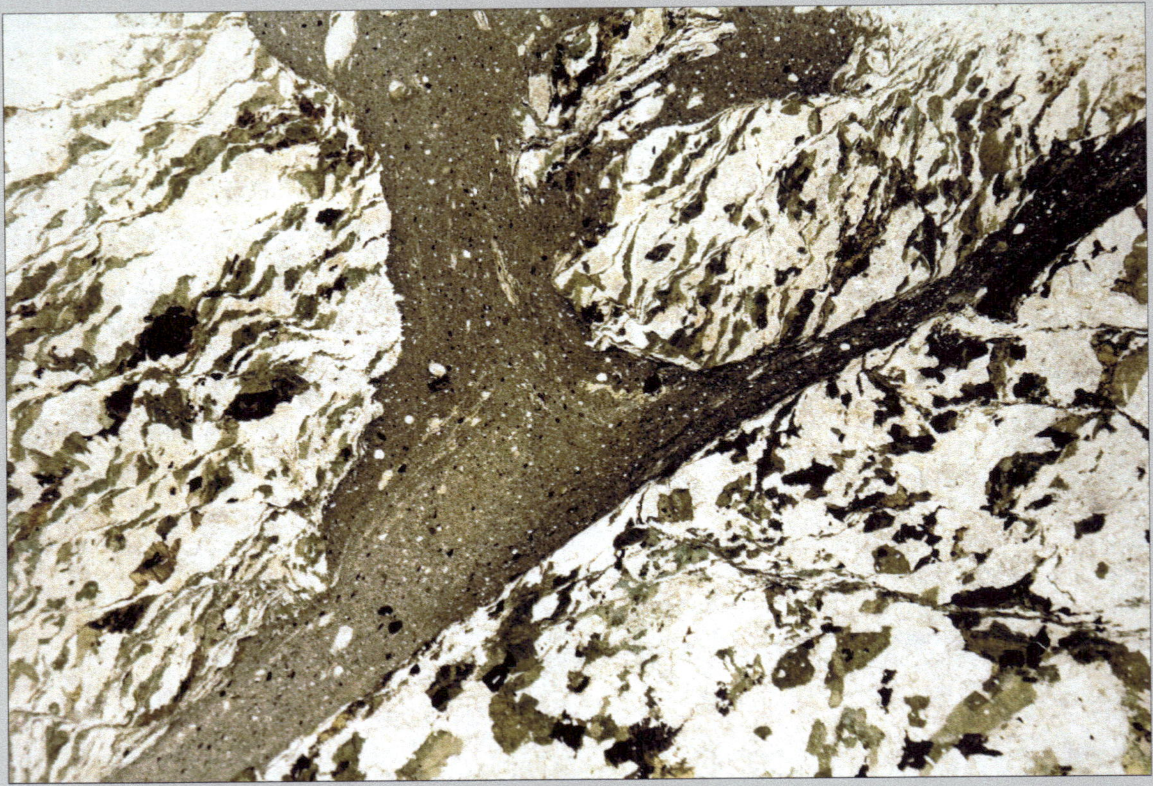

Fig. 3.5 Pseudotachylyte vein in deformed granodiorite. A main fault vein or generation surface cuts the host rock from upper right to lower left with an injection vein propagating upwards. The dark pseudotachylyte matrix contains somewhat rounded inclusions, mainly of quartz and opaque minerals. The boundaries between the pseudotachylyte and the wall rock are sharp. Torres de Paine, southern Chile. Width of view 20 mm. PPL.

Fig. 3.6 As Fig. 3.5. Width of view 20 mm. CPL.

Fig. 3.7 Detail of Fig. 3.5 showing the lobate shape of the contact of the injection vein of pseudotachylyte with the granodioritic wall rock. Rounded inclusions of quartz and opaque material lie in the matrix. Width of view 3 mm. PPL.

Fig. 3.8 Pseudotachylyte vein in a granitic host rock. The straight lower contact corresponds probably to the generation surface. Note several minor injection veins (e.g. at lower right) and the flow structure. Leiden collection. Width of view 20 mm. CPL.

Fig. 3.9 Typical contact of a pseudotachylyte vein (upper part) and a granitic host rock (lower part). Note the orthogonal injection vein below the center and the rounded inclusions within the pseudotachylyte. Leiden collection. Width of view 4 mm. PPL.

Fig. 3.10 Detail of pseudotachylyte vein showing typical lobate and circular "devitrification" structures around gas bubbles in a fine-grained matrix. The gas bubbles are now filled with opaque material and carbonate. Leiden collection. Width of view 2.5 mm. PPL.

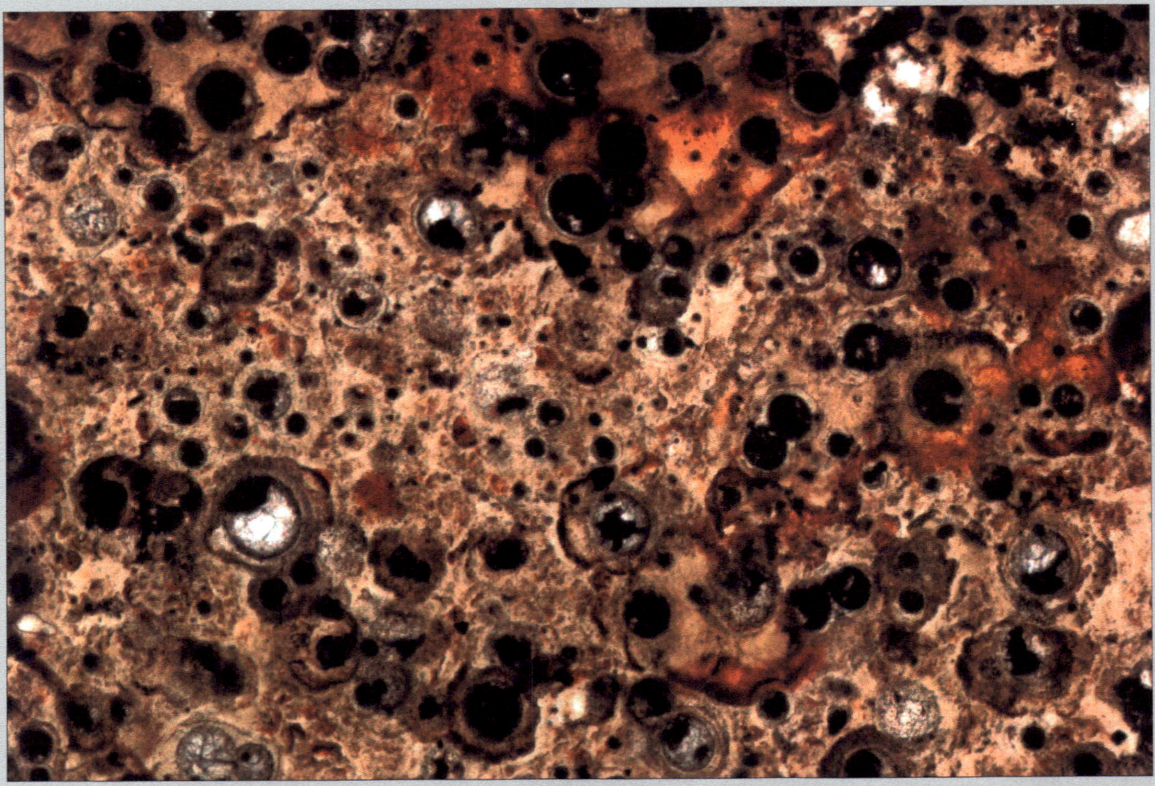

Fig. 3.11 Pseudotachylyte vein (black) in a host rock of cataclastic protomylonitic granite. The pseudotachylyte vein has sharp contacts with cataclasite in the center and along the lower contact. The cataclasite was probaly the precursor rock from which the pseudotachylyte formed. Leiden collection. Width of view 20 mm. PPL.

Fig. 3.12 As Fig. 3.11. Width of view 20 mm. CPL.

Fig. 3.13 Pseudotachylyte vein (dark) in a host rock of protomylonitic cataclastic orthogneiss. An injection vein intrudes upwards, left of the center. Many minor faults can be observed in the host rock. Inclusions in the pseudotachylyte vein are stretched upper left to lower right (comparable to oblique foliation), indicative of ductile deformation superposed on the pseudotachylyte formation, probably with a sinistral sense of motion. Roraima, northern Brazil. Width of view 13 mm. PPL.

Fig. 3.14 As Fig. 3.13. Width of view 13 mm. CPL.

Fig. 3.15 Detail of Fig. 3.13, showing how the injection vein (above the center) is connected to the main pseudotachylyte vein. The foliation induced by ductile deformation in the pseudotachylyte vein is horizontal in this view. Width of view 4 mm. PPL.

Fig. 3.16 Pseudotachylyte vein (dark, lower part) in a host rock of mylonite (upper part) to protomylonite (lower part) with many minor faults. Minor injection veins are common, especially along the lower contact of the pseudotachylyte vein. Ductile deformation within the pseudotachylyte vein has induced an "oblique foliation" by deformation of light colored inclusions. This "oblique foliation" and asymmetric folds in the host rock are indicative of dextral sense of shear. The folded and faulted dark bands in the upper part of the photomicrograph are ultramylonitic. Roraima, northern Brazil. Width of view 17 mm. PPL

Fig. 3.17 As Fig. 3.16. Width of view 17 mm. CPL

Fig. 3.18 Typical contact of a pseudotachylyte vein (dark upper part) in a granitic host rock. Note the inclusions in the pseudotachylyte, the sharpness of the contact, the presence of minor injection veins and the fractures in the host rock. Leiden collection. Width of view 4 mm. PPL.

Fig. 3.19 Pseudotachylyte (black domain in the upper part) veins in a host rock of mylonite derived from a felsic granulite. The porphyroclasts in the mylonite are composed of feldspar and orthopyroxene. The matrix is rich in biotite. Elongated inclusions in the pseudotachylyte are the result of minor ductile deformation after the solidification of the pseudotachylyte. Note the irregular shape of the pseudotachylyte vein and some isolated patches (right of the center) that may be linked to the main pseudotachylyte body in the third dimension. A minor fault with apparent sinistral offset is present left of the center. No general sense of shear can be deduced from this photomicrograph. Mount West, Western Australia. Width of view 18 mm. PPL.

Fig. 3.20 As Fig. 3.19. Width of view 18 mm. CPL.

Fig. 3.21 Pseudotachylyte vein (black, lower left to upper right) with relatively large and abundant inclusions of host rock. The host rock is a low-grade mylonite rich in biotite and garnet below the vein, with few orthopyroxene porphyroclasts, and rich in quartz and feldspar above the pseudotachylyte vein. All large grey crystals in the upper part, with undulose extinction, are feldspars. Mount West, Western Australia. Width of view 22 mm. CPL.

Fig. 3.22 Detail of Fig. 3.21, showing flow structure in the pseudotachylyte. Width of view 4 mm. PPL.

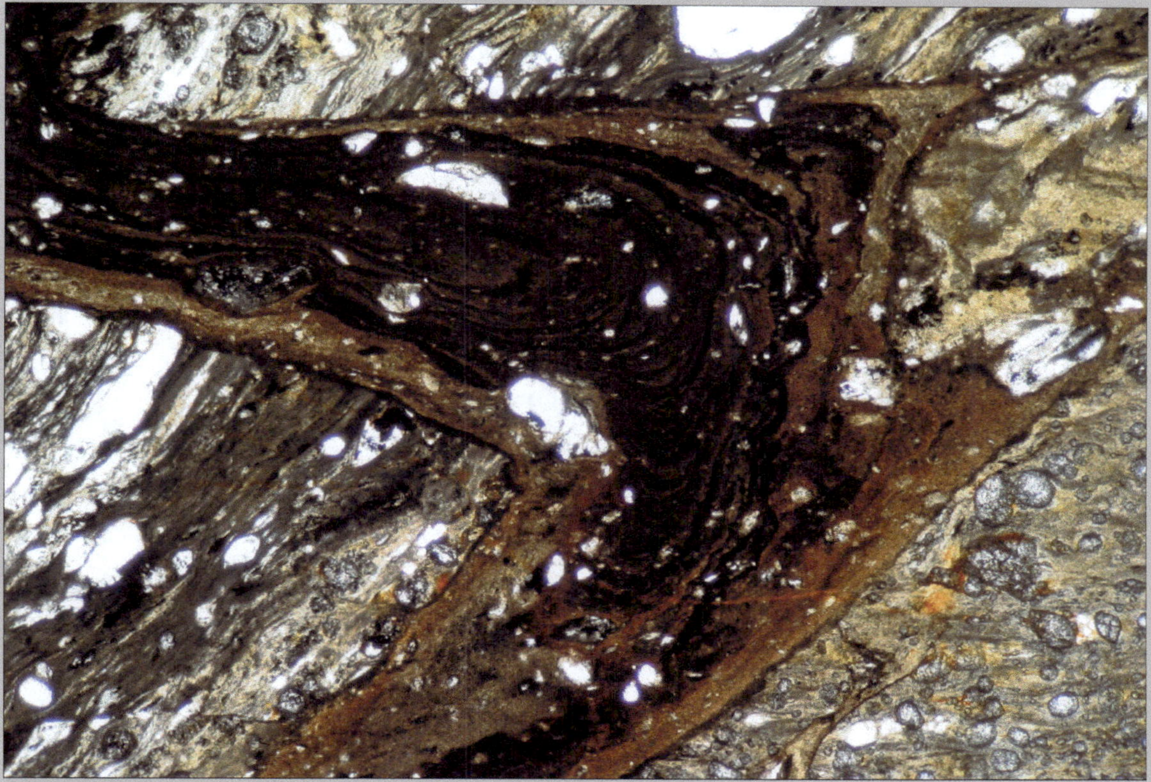

Fig. 3.23 Pseudotachylyte vein (black) in a host rock of mylonitised granulite with porphyroclasts of plagioclase, garnet and orthopyroxene embedded in a biotite-rich matrix. The pseudotachylyte vein has sharp contacts and an irregular shape. Some parts of the matrix (upper right) show cataclastic deformation in addition to mylonitisation. Mount West, Western Australia. Width of view 20 mm. PPL.

Fig. 3.24 As Fig. 3.23. Width of view 20 mm. CPL.

Fig. 3.25 Pseudotachylyte vein (dark upper part) in folded schist. Note the embayment at upper left with the internal flow structure typical of pseudotachylytes. Leiden collection. Width of view 14 mm. PPL.

Fig. 3.26 As Fig. 3.25. Width of view 14 mm. CPL.

Fig. 3.27 Detail of Fig. 3.25, showing the flow structure in the embayment of the pseudotachylyte vein. Width of view 3 mm. PPL.

Fig. 3.28 Pseudotachylyte vein in protomylonitic to cataclastic granitic host rock. Note the injection vein intruding downwards from the lower contact. Many inclusions are relatively angular. A vague flow structure is visible in the vein suggesting flow from left to right. Leiden collection. Width of view 25 mm. PPL.

Fig. 3.29 As Fig. 3.28. Width of view 25 mm. CPL.

Chapter 4 | Low-Grade Mylonites

4 Low-Grade Mylonites

The temperature range for these mylonites is thought to be roughly between 250 and 500 °C. There is a gradual transition between cataclasites and low-grade mylonites. Whereas many feldspar porphyroclasts in low-grade mylonites still show fracturing by cataclasis, the quartz is usually deformed by crystal-plastic processes as shown by its change in shape and by undulose extinction. At increasing temperature bulging recrystallisation starts to manifest itself along the lobate contacts and eventually recrystallisation by subgrain rotation takes over (Chapter 10).

The main distinguishing feature of low- and medium-grade mylonites is the size to which recrystallised quartz has grown. In low-grade mylonites this size is smaller than 50 micrometers whereas in medium-grade ones it is larger. Although felspar is mostly deformed by fracturing, crystal-plastic deformation is evident by undulose extinction and comminution into tiny new grains.

Biotite is usually reduced to very small grains in the matrix, but muscovite tends to resist deformation and may develop into porphyroclastic "fish", especially when surrounded by quartz. In micas and feldspars deformed at low grade, it is usually hard to determine if grain comminution is due to cataclasis or recystallisation.

Another characteristic feature of low-grade mylonites is the frequent presence of sharp transitions between protomylonites, mylonites and ultramylonites, reflecting strong gradients of strain intensity. Asymmetric structures that can be used as shear sense indicators (Chapter 9) are best developed in low-grade mylonites; these are less common in cataclasites and high-grade mylonites.

It is important to realise that, when estimating metamorphic grade during mylonitisation from the microstructure, we actually judge the conditions during the final deformation stage. Earlier stages may be overprinted and not recognisable any more. In some cases this superposition may be only partial, leaving evidence for activity at different temperatures (Chapter 9.8).

Fig. 4.1 A very low-grade mylonite, transitional to cataclasite, derived from hornblende syenite. The large K-feldspar porphyroclast is broken into many fragments that are progressively incorporated in the foliated matrix. No recrystallisation is visible. A mylonitic foliation is developing in the upper and lower parts of the photomicrograph. No sense of shear is apparent in this photomicrograph. Twijfelfontein, Kaokoveld, NW Namibia. Width of view 14 mm. CPL.

Fig. 4.2 Example of a mosaic-fragmented feldspar porphyroclast in a low-grade mylonite derived from granite. Sinistral sense of shear can be deduced from the sinistral offset (syntaxial) along faults slightly inclined to the left and dextral offset along faults steeply inclined to the right (antitaxial). St. Barthélemy Massif, French Pyrenees. Width of view 4 mm. CPL.

Fig. 4.3 Low-grade mylonite derived from granite. Notice the narrow transition between protomylonite in the upper part and ultramylonite near the lower edge of the photograph, characteristic for low-grade mylonites. No recrystallisation is apparent at this magnification. The small angle between the more steeply inclined mylonitic foliation in the upper part and the more flat lying one in the lower part indicates dextral sense of movement (C/S fabric). St. Barthélemy Massif, French Pyrenees. Width of view 12 mm. CPL.

Fig. 4.4 Detail from the same thin section as Fig. 4.3, showing contrasting behaviour of K-feldspar and quartz. The K-feldspar (above center) is broken into a dozen slices dislocated in an en- echelon pattern. They show minor crystalplastic deformation. The lower part of the photomicrograph is mainly composed of one large quartz grain with strong undulose extinction, indicating considerable crystalplastic deformation and minor recrystallisation into very small new grains, along high strain zones that nucleate from intracrystalline faults in the feldspar. Width of view 4 mm. CPL.

Fig. 4.5 Low-grade mylonite derived from granite. Quartz is strongly deformed by crystal-plastic deformation and does not show any recrystallisation. The C/S fabric in the upper left part of the photomicrograph and the inclined shear band (C´) in the lower right part are both indicative of sinistral sense of shear. A feldspar porphyroclast in the central upper part is broken into small fragments. Oropa, Italian Alps. Width of view 17 mm. PPL.

Fig. 4.6 As Fig. 4.5. Width of view 17 mm. CPL.

Fig. 4.7 Detail of the low-grade mylonite of Fig. 4.6. The quartz-rich bands show strong undulose extinction due to crystalplastic deformation, without visible recrystallisation. The mica-rich matrix is mainly composed of very fine-grained biotite. Oblique shear bands (C´) from upper right to lower left indicate sinistral sense of shear. Width of view 6 mm. CPL.

Fig. 4.8 Other detail of the low-grade mylonite shown in Fig. 4.6. Individual quartz grains are strongly deformed and draped around the more resistant K-feldspar porphyroclast in the center. Note the undulose extinction both in the quartz and in the K-feldspar. Stair stepping around the porphyroclast indicates sinistral sense of shear. Width of view 3 mm. CPL.

Fig. 4.9 Mylonite derived from a schist, containing biotite, muscovite, quartz, garnet and kyanite. Although the schist is of amphibolite facies, the mylonitisation, or at least the last mylonitisation, occurred at lower greenschist facies conditions, as indicated by the quartz ribbons with strong undulose extinction (Fig. 4.10) and only partial recrystallisation by bulging (see Fig. 4.11). Dextral sense of shear is suggested by oblique fabric in the quartz ribbons, some mica fish and incipient shear bands (C´) in the lower biotite and garnet-rich band. Marsfjällen, Västerbotten, Sweden. Width of view 16 mm. PPL.

Fig. 4.10 As Fig. 4.9. Width of view 16 mm. CPL.

Fig. 4.11 Detail of Fig. 4.10, showing incipient recrystallisation by bulging in the quartz vein above the center. The lobate contacts of the deformed quartz grains define an oblique foliation consistent with dextral shear sense. Width of view 3 mm. CPL.

Fig. 4.12 Mylonite derived from paragneiss with garnet, kyanite, biotite and K-feldspar. Note the strong deviation of the mylonitic foliation around the garnet porphyroblast at right. The yellow grains in Fig. 4.13 are kyanite crystals. Note that the quartz grains below the garnet were protected from deformation, probably by extension of the garnet below or above the thin section (in 3D). No sense of shear is apparent in this view. Marsfjällen, Västerbotten, Sweden. Width of view 16 mm. PPL.

Fig. 4.13 As Fig. 4.12. Width of view 16 mm. CPL.

Fig. 4.14 Detail of Fig. 4.13 to show deformed quartz with undulose extinction and lobate contacts. Note incipient recrystalisation by bulging, indicative of low-grade metamorphic conditions during mylonitisation. Width of view 3 mm. CPL.

Fig. 4.15 Low-grade mylonite derived from a granite. Feldspar porphyroclasts, some with undulose extinction are surrounded by a quartz-rich matrix. Notice limited recrystallisation in quartz (Fig. 4.17), mainly by bulging. Sense of shear is dextral as indicated by the C/S fabric left of the center. Mount West, Australia. Width of view 18 mm. PPL.

Fig. 4.16 As Fig. 4.15. Width of view 18 mm. CPL.

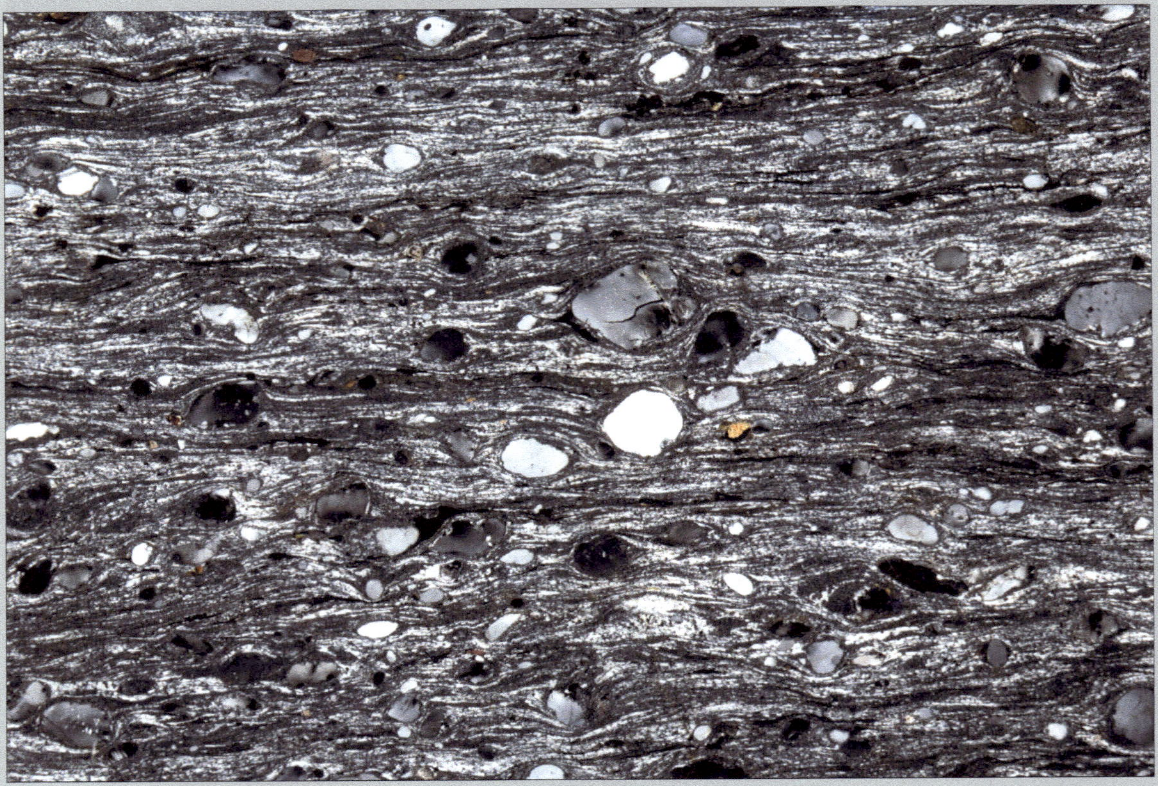

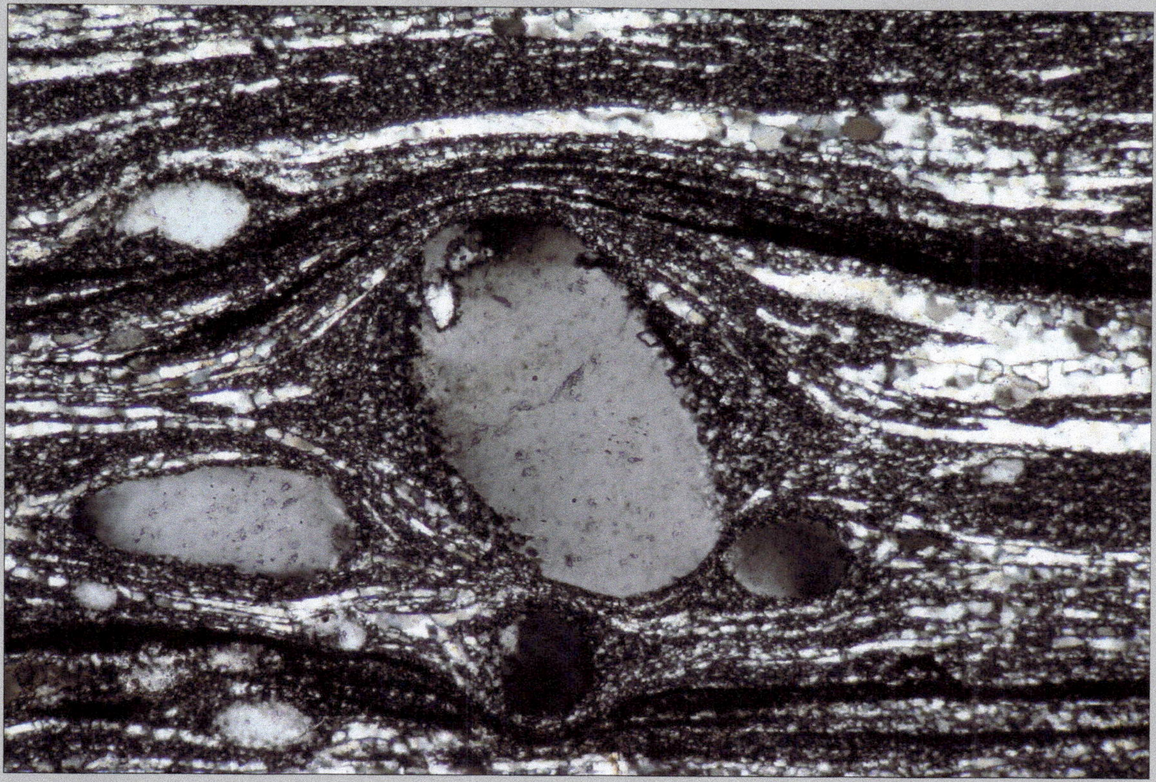

Fig. 4.17 Detail of Fig. 4.16, showing a porphyroclast of K-feldspar with undulose extinction and apparent recrystallisation to very small grains mainly concentrated in the strain shadows. Quartz (light colored veins e.g. upper right) shows minor recrystallisation by bulging, consistent with low metamorphic conditions. Width of view 1.25 mm. CPL.

Fig. 4.18 Low-grade mylonitised granite. Most porphyroclasts show undulose extinction (Fig. 4.19). Biotite-rich seams in the matrix were strongly sheared and became C planes, separating lenses with less strain and preserved S planes (C/S fabric). The C/S fabric indicates dextral shear. Asymmetric folds at lower left support this interpretation. Western Australia. Width of view 16 mm. PPL.

Fig. 4.19 As Fig. 4.18. Width of view 16 mm. CPL.

Fig. 4.20 Detail of Fig. 4.19, showing a strongly deformed porphyroclast of plagioclase with accentuated undulose extinction and minor recrystallisation to a very fine-grained aggregate of new grains, especially along the upper contact. Dextral sense of shear is indicated by the stair stepping from left to right across the porphyroclast. Quartz (upper right part) is strongly deformed with only minor recrystallisation, pointing to low temperature conditions. Width of view 3 mm. CPL.

Fig. 4.21 Mylonitised granulite with porphyroclasts of plagioclase, orthopyroxene and amphibole embedded in a dark matrix. A sinistral sense of shear can be deduced from a large delta-type plagioclase porphyroclast above the center, with strong undulose extinction (Fig. 4.22). Low metamorphic conditions are indicated by the highly elongated deformed quartz grains in the lower part of the photo (see also Fig. 4.23). Western Australia. Width of view 12 mm. PPL.

Fig. 4.22 As Fig. 4.21. Width of view 12 mm. CPL.

Fig. 4.23 Detail of Fig. 4.22. Note the contrasting behaviour of quartz and plagioclase: the plagioclase porphyroclasts are deformed by crystal-plastic deformation showing undulose extinction and folded twin lamellae, whereas the quartz (lower right) exhibits lobate contacts with incipient bulging recrystallisation, characteristic for low-grade metamorphic conditions. The naked yellowish porphyroclasts are pyroxene.
Width of view 1.25 mm. CPL.

Fig. 4.24 Mylonitised pegmatitic granite. Feldspar porphyroclasts are strongly deformed by crystal-plastic deformation. Quartz is even stronger deformed to highly undulose ribbons that are to a large extent recrystallised to small granoblastic new grains. Sense of shear is not obvious in this photomicrograph (it is dextral).
Cadaquez, NE Spain. Width of view 22 mm. CPL.

Fig. 4.25 Detail of Fig. 4.24, showing a strongly deformed plagioclase porphyroclast, surrounded by fine-grained recrystallised quartz. Width of view 11 mm. CPL.

Fig. 4.26 Close-up of the plagioclase porphyroclast shown in Fig. 4.25. Notice internal kinking and folding of the plagioclase crystal, without recrystallisation. Width of view 4 mm. CPL.

Fig. 4.27 Detail of Fig. 4.24, showing undulose extinction in a large plagioclase porphyroclast that seems to be boudinaged. The quartz is recrystallised to a polygonal granoblastic aggregate composed of small new grains. Small mica fish in the lower part indicate dextral sense of shear. Width of view 3.5 mm. CPL.

Fig. 4.28 This is also a detail of Fig. 4.24, showing an oblique foliation of mica in the upper part, indicating dextral sense of shear. The quartz in the vein close to the bottom is partially recrystallised by bulging, characteristic for low temperature conditions. Width of view 2 mm. CPL.

Fig. 4.29 Low-grade mylonite derived from granite. Porphyroclasts are composed of feldspar. The matrix contains fine-grained biotite and quartz. Sense of shear is sinistral as can be deduced from stair stepping to the left across the porphyroclasts. Deflection of the mylonitic foliation around the porphyroclasts (Fig. 4.30) causes the false impression of a late crenulation with upper left – lower right trending axial plane. Another trace of the axial planes of incipient extensional crenulation, trending upper right – lower left (visible in the upper ultramylonite band) is interpreted as related to C´ shear bands. South Island, New Zealand. Width of view 18 mm. PPL.

Fig. 4.30 As Fig. 4.29. Width of view 18 mm. CPL.

Fig. 4.31 Detail of Fig. 4.30 showing partial recrystallisation of quartz in the tip of a large K-feldspar porphyroclast. The biotite-rich matrix contains muscovite fish (pink); a large biotite crystal (green) included in the K-feldspar porphyroclast is protected from the deformation. Width of view 3mm. CPL.

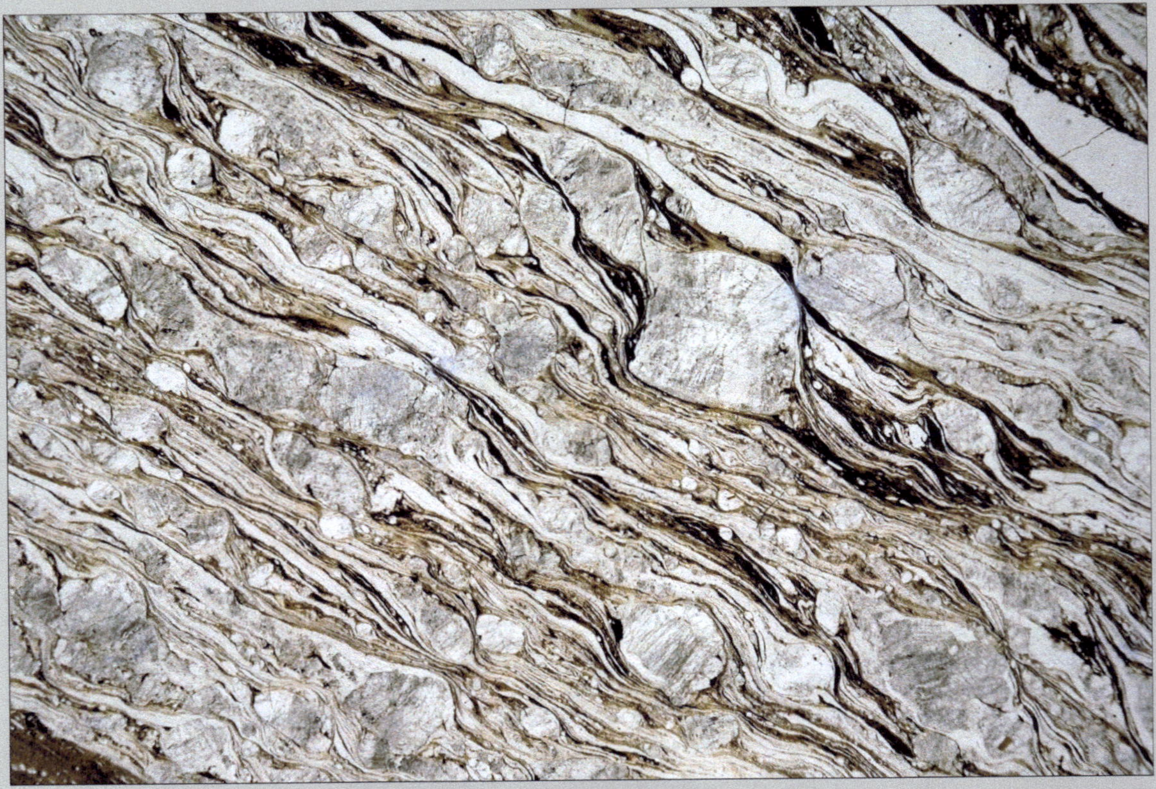

Fig. 4.32 Low-grade mylonite derived from granite. The porphyroclasts are composed of feldspar. A sinistral sense of shear can be inferred from minor asymmetric folds (upper right) and stair stepping to the left. South Island, New Zealand. Width of view 30 mm. PPL.

Fig. 4.33 As Fig. 4.32. Width of view 30 mm. CPL.

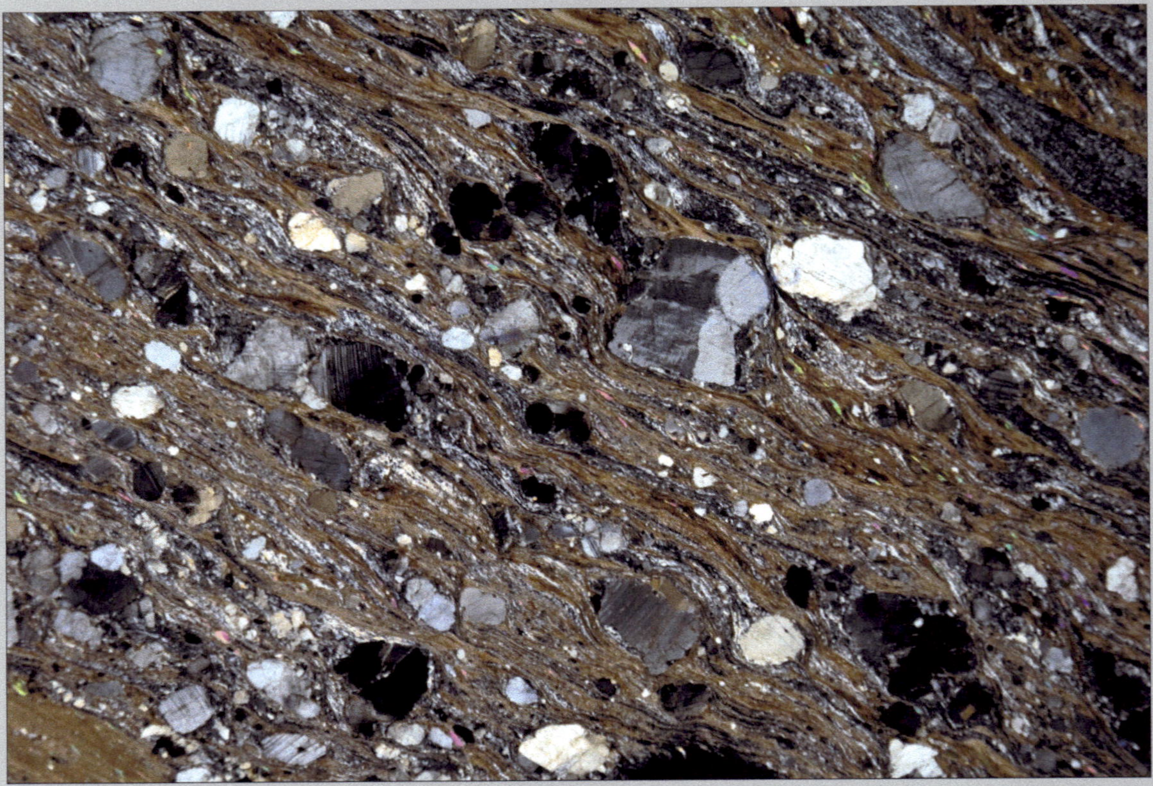

Fig. 4.34 Detail of Fig. 4.33; the central porphyroclast of plagioclase is broken and offset along a sinistral upper right – lower left trending fault, syntaxial to the movement in the matrix. The quartz in the upper horizontal band is stretched and partially recrystallised, forming an obique fabric also indicating sinistral sense of shear. The mica fish at lower left is yet another shear sense indicator. Width of view 3 mm. CPL.

Fig. 4.35 Low-grade mylonite to ultramylomite derived from orthogneiss. The sheared-out quartz lenses with limited recrystallisation (Figs. 4.36 and 4.37) and the broken feldspar fragments are diagnostic for low temperature deformation. A dextral sense of shear is indicated by several asymmetric minor folds (upper part, left of center), a vague C/S fabric and syntaxial offset of feldspar fragments (Fig. 4.37). Southern Minas Gerais State, SE Brazil. Width of view 12 mm. PPL.

Fig. 4.36 As Fig. 4.35. Width of view 12 mm. CPL.

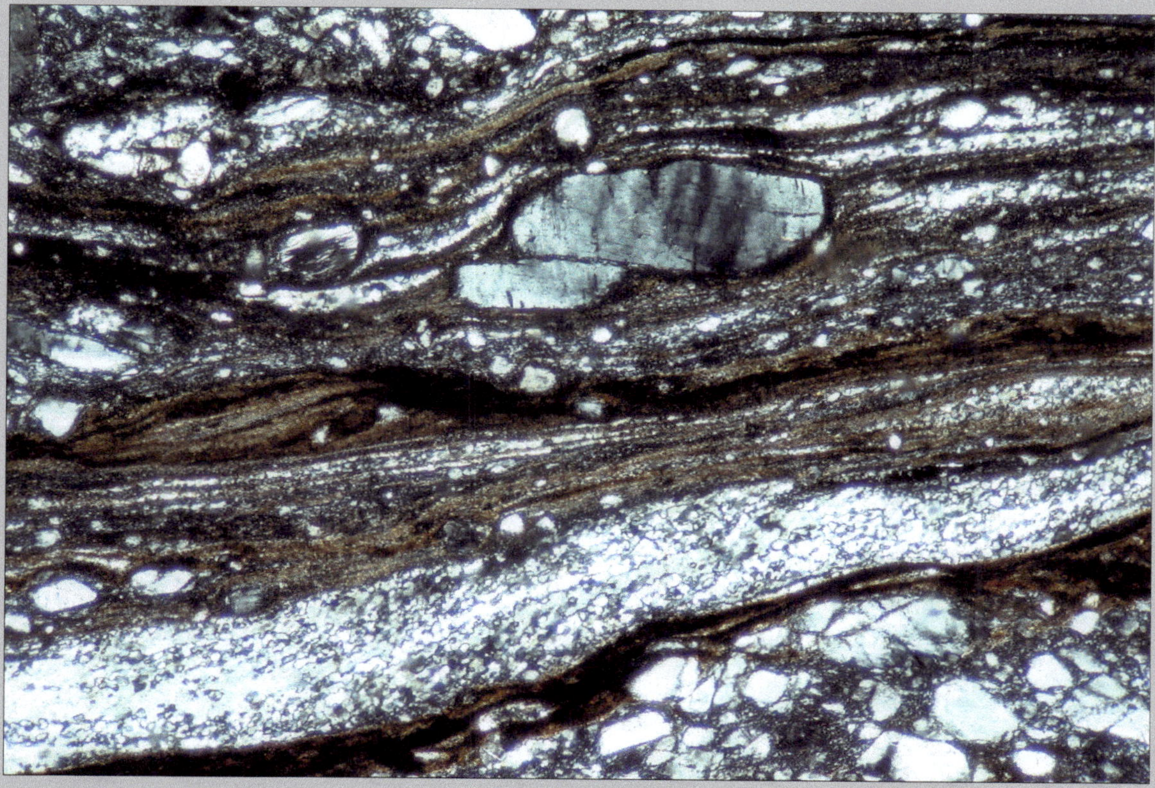

Fig. 4.37 This detail of Fig. 4.36 shows dextral offset within the feldspar porphyroclast, syntaxial to the general movement in the matrix. The quartz band below the center is partially recrystallised to small new grains and the feldspar below this band is broken to angular fragments, apparently by brittle deformation. Width of view 2 mm. CPL.

Fig. 4.38 Close-up of the quartz band shown in Fig. 4.37 to show partial recrystallisation to very small new grains by bulging and minor subgrain rotation. The oblique fabric in quartz is consistent with dextral sense of shear. Width of view 0.75 mm. CPL.

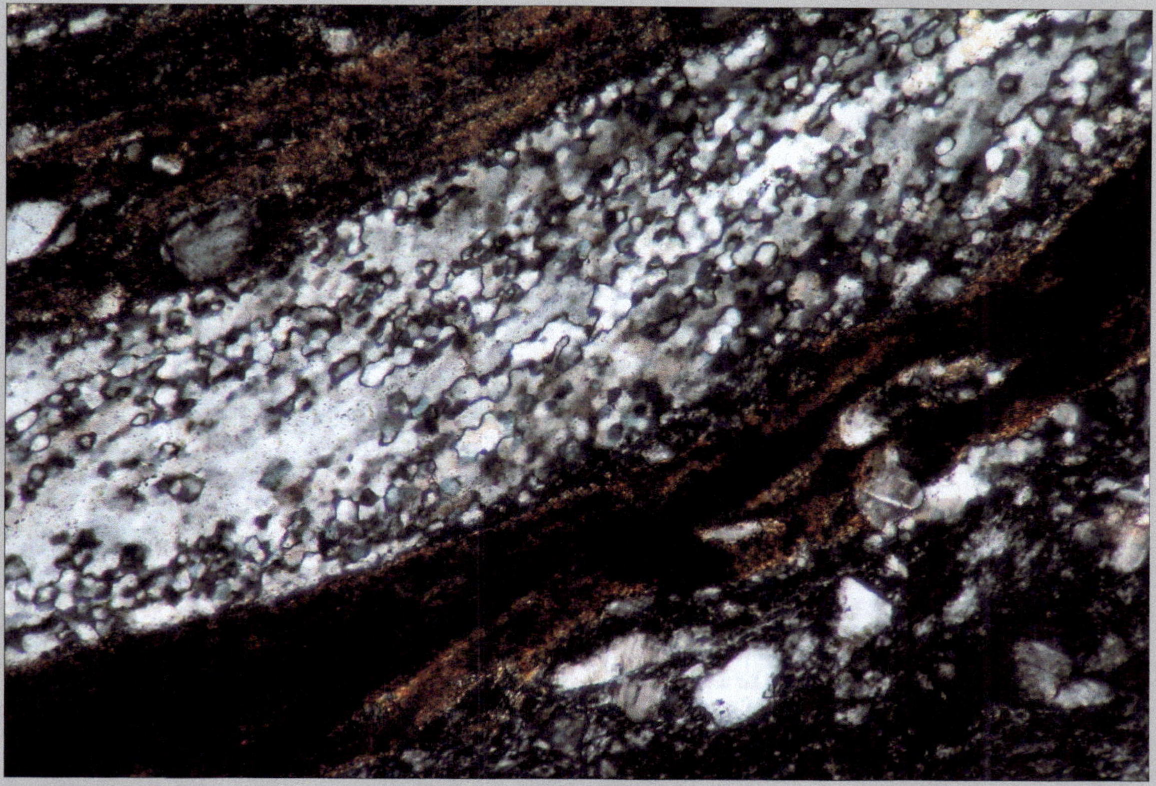

Fig. 4.39 Low-grade mylonite from the same outcrop as Figs. 4.35 - 4.38. Notice the strong contrast between the inclined protomylonite in the lower part, with little matrix, and the dark ultramylonite bands in the upper part. This strong partitioning of strain intensity is common in low-grade mylonites. A dextral sense of shear is indicated by the angle between the less deformed lower part (S plane) and the ultramylonite bands (C planes). Width of view 18 mm. PPL.

Fig. 4.40 As Fig. 4.39. Width of view 18 mm. CPL.

Fig. 4.41 Detail of Fig. 4.40. showing sinistral offset of feldspar fragments along steeply inclined faults, antitaxial with respect to the main, dextral sense of shear in the matrix. Width of view 3 mm. CPL.

Fig. 4.42 Low to medium-grade mylonite with porphyroclasts of muscovite and feldspar in a matrix of very fine-grained biotite and quartz, derived from a paragneiss. The oblique mica fish indicate sinistral sense of shear. Note the strong contrast between the sheared-out and partially recrystallised quartz and feldspar that shows evidence of both brittle and ductile deformation. Pernambuco, NE Brazil. Width of view 4 mm. CPL.

Fig. 4.43 Low-grade mylonite derived from garnet-bearing granite. The central porphyroclast is garnet; the other white porphyroclasts are feldspar and the matrix is composed of fine-grained biotite and quartz. This photomicrograph shows a clear contrast between the brittle behaviour of the garnet and the ductile behaviour of the matrix. Internal faults in the garnet show apparent dextral dislocation ("domino-type fragmented porphyroclasts", see section 9.7) but the shear sense in the mylonite is sinistral as can be inferred from the C/S fabric in the lower part and stair stepping across the garnet. Pernambuco, NE Brazil. Width of view 4 mm. PPL.

Fig. 4.44 This photomicrograph shows another view of the same thin section as Fig. 4.43 with an elongated lens of garnet fragments, just below the center. This lens results from cataclasis in the garnet and subsequent extension along with the matrix deformation. The large whitish porphyroclasts are feldspar and some lens-shaped ones are muscovite. A sinistral shear sense can be inferred from stair stepping across asymmetric porphyroclasts and mica fish inclined to the right (upper right). Width of view 4 mm. PPL.

Chapter 5 | Medium-Grade Mylonites

5 Medium-Grade Mylonites

The temperature range for the formation of this group of mylonites is approximately 500 to 650 °C. In medium-grade mylonites quartz is usually fully recrystallised, mainly by subgrain rotation, and has grown to a polygonal crystalloblastic fabric of strain free grains with an average grain size exceeding about 50 micrometers.

Feldspar porphyroclasts may show undulose extinction and core-mantle structures due to partial recrystallisation. Fracturing is not common. Strong contrasts between protomylonites, mylonites and ultramylonites are rare and gradual transitions to non-mylonitic rocks are common. Asymmetric structures that can be used as shear sense indicators are not as frequent as in low-grade mylonites.

Fig. 5.1 Mylonite derived from kyanite bearing paragneiss. The porphyroclasts are plagioclase and kyanite (coloured crystals) and the matrix is composed of small biotite, muscovite and quartz. Medium-grade conditions during and after mylonitisation can be inferred from the relatively large grain size of recrystallised quartz. Sense of shear is sinistral, indicated by several kyanite fish inclined to the right, and a C´ shear band in the upper central part. Caxambu, southern Minas Gerais State, SE Brazil. Width of view 10 mm. CPL.

Fig. 5.2 Mylonite derived from kyanite bearing paragneiss. At lower left, garnet (black) contains quartz inclusions. A C/S fabric, with S inclined to the right and C subhorizontal is well developed and indicates sinistral sense of shear. Notice the well-recrystallised quartz, grown to a moderate grain size. Caxambu, southern Minas Gerais State, SE Brazil. Width of view 10 mm. CPL.

Fig. 5.3 Medium-grade mylonite derived from granite. The large porphyroclasts are K-feldspar, embedded in a matrix of small recrystallised K-feldspar, fine-grained biotite and quartz. Medium-grade conditions can be inferred from the advanced recrystallisation of K-feldspar along the rims of porphyroclasts (Figs. 5.5 and 5.6). A dextral sense of shear is indicated by stair stepping to the right across the porphyroclasts and an incipient C/S fabric with S planes inclined slightly to the left and C planes inclined slightly to the right. Western Australia. Width of view 18 mm. PPL.

Fig. 5.4 As Fig. 5.3. Width of view 18 mm. CPL.

Fig. 5.5 Detail of Fig. 5.4, showing a K-feldspar porphyroclast with undulose extinction and subgrain development. The porphyroclast is surrounded by small K-feldspar grains that result from recrystallisation, probably mainly by subgrain rotation. Width of view 2 mm. CPL.

Fig. 5.6 Close-up of the rim of the same K-feldspar porphyroclast as Fig. 5.5, to show the transition between deformed porphyroclast with subgrains and the crystalloblastic fabric of recrystallised K-feldspar. The horizontal grains along the bottom are quartz. Width of view 0.75 mm. CPL.

Fig. 5.7 Mylonite derived from a mica-rich granite. Porphyroclasts are of feldspar surrounded by a biotite and quartz-rich matrix. Biotite fish (green) are inclined towards the right indicating sinistral sense of shear. A C/S fabric has S planes inclined to the right and C planes subhorizontal or slightly inclined to the left. The size of recrystallised quartz is indicative for low- to medium-grade conditions during mylonitisation. St. Barthélemy Massif, French Pyrenees. Width of view 12 mm. CPL.

Fig. 5.8 Mylonitic schist with foliation fish composed of muscovite enveloping garnet and staurolite grains. The matrix is mainly composed of quartz, biotite and muscovite and resembles an ordinary schist rather than a mylonite, showing static recrystallisation. The mylonitic structure is apparent from the curvature of the foliation fish that indicates dextral sense of shear. The fabric can also be described as a C/S fabric with subhorizontal S planes and C planes inclined to the right. Santa Rita do Ibitipoca, southern Minas Gerais State, SE Brazil. Width of view 16 mm. CPL.

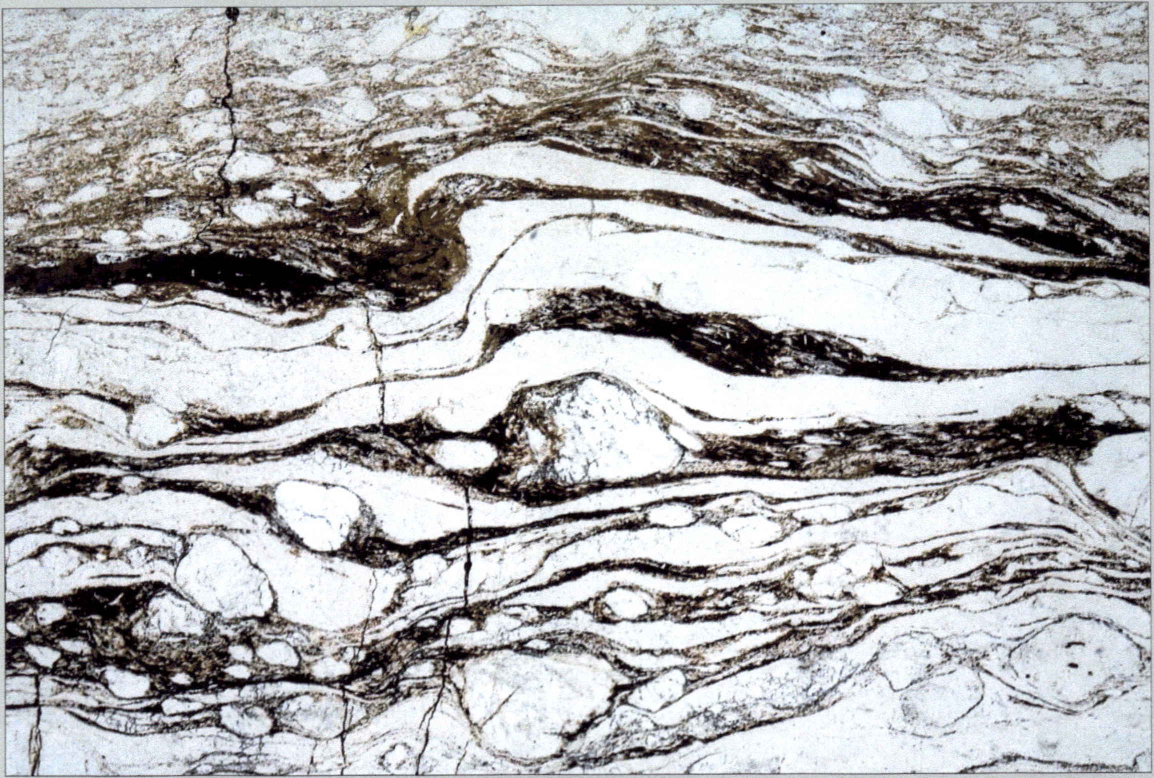

Fig. 5.9 Low- to medium-grade mylonite derived from paragneiss. The porphyroclasts are feldspar embedded in a matrix of quartz and mica. A sinistral sense of shear can be inferred from asymmetric folds (above center) and inclined mineral fish at upper right. The well-recrystallised quartz, grown to medium grain size, is indicative of the metamorphic conditions. Cadaquez, NE Spain. Width of view 16 mm. PPL.

Fig. 5.10 As Fig. 5.9. Width of view 16 mm. CPL.

Fig. 5.11 Medium-grade mylonite derived from a kyanite K-feldspar paragneiss. The porphyroclasts are K-feldspar and kyanite, surrounded by a matrix of biotite and quartz. The large K-feldspar to the right of the center is partially recrystallised especially along high strain zones. Sense of shear cannot be established with certainty from this image. Marsfjället, Västerbotten, Sweden. Width of view 14 mm. CPL.

Fig. 5.12 Detail of Fig. 5.11, showing advanced recrystallisation around the K-feldspar porphyroclasts. The transition between the strongly deformed K-feldspar to a fine-grained mosaic of new grains, probably mainly recrystallised by subgrain rotation, can be observed on the right hand side of the large porphyroclast. The K-feldspar at lower right shows curved perthitic exsolution due to crystal-plastic deformation. At lower left the horizontal quartz-biotite mylonitic matrix foliation is just visible. Width of view 5 mm. CPL.

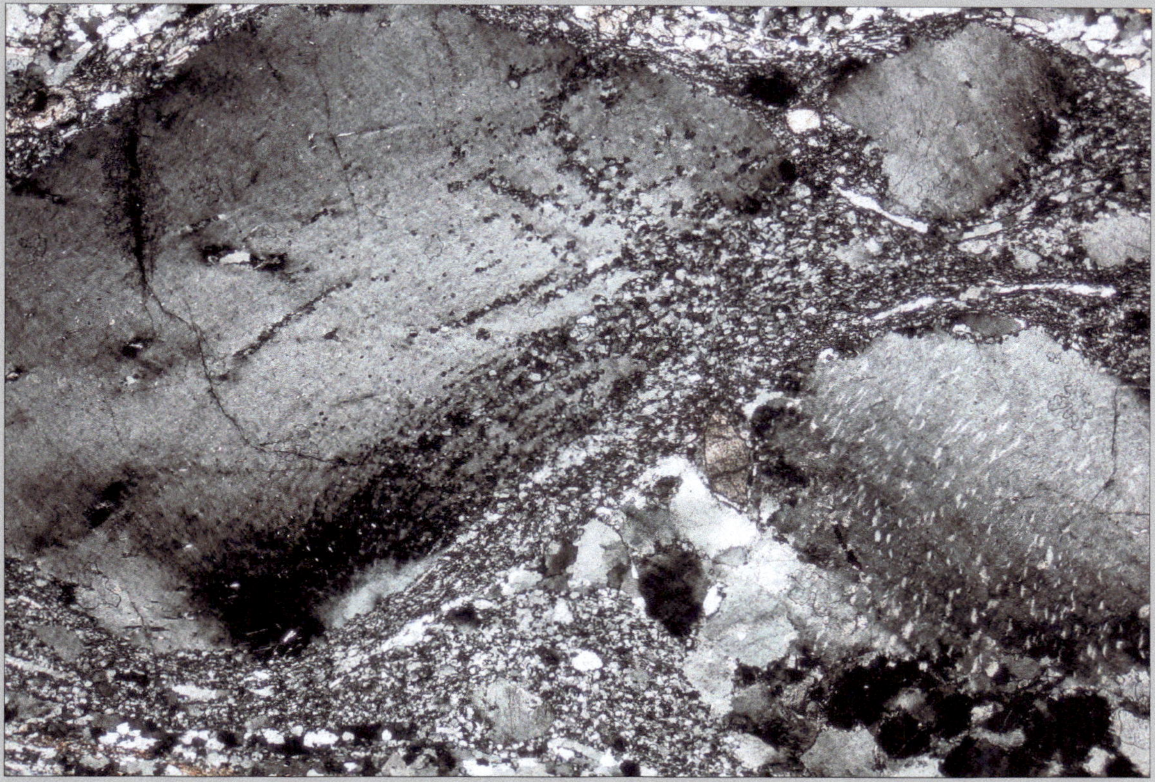

Fig. 5.13 Detail from the same sample as Fig. 5.11, showing a large kyanite crystal covering most of the photomicrograph with internal deformation by kinking and folding. At lower and middle right intergrown biotite can be observed and at upper left a strain free recrystallised quartz grain is visible. Width of view 1.2 mm. CPL.

Fig. 5.14 Strongly kinked kyanite crystals indicating intense deformation, surrounded by recrystallised granoblastic quartz, indicative of medium temperature (upper greenschist to lower amphibolite facies). Lavras, southern Minas Gerais State, SE Brazil. Width of view 3 mm. CPL.

Fig. 5.15 Low- to medium-grade mylonite derived from garnet mica schist. Most coloured grains are muscovite fish, at left are two garnet crystals (black in Fig. 5.16) and in the upper part are some plagioclase porphyroclasts. Quartz is well recrystallised. Sense of shear is sinistral, indicated by the inclination to the right of the mica fish. Caxambu, southern Minas Gerais State, SE Brazil. Width of view 25 mm. PPL.

Fig. 5.16 As Fig. 5.15. Width of view 25 mm. CPL.

Fig. 5.17 Detail of Fig. 5.16 that shows strong undulose extinction of muscovite fish (coloured grains) and a granoblastic fabric in quartz. The strain-free statically recrystallised quartz fabric shows that the temperature at the end of mylonitisation was at medium metamorphic grade, probably close to the transition between greenschist and amphibolite facies. Sense of shear is sinistral, indicated by mica fish inclined to the right and by an upper right – lower left C′ type shear band in the upper part. Width of view 3 mm. CPL.

Fig. 5.18 Medium-grade micaceous quartzite with a recrystallised foliation fish composed of muscovite. Most muscovite grains are recrystallised to strain-free new grains without undulose extinction. Quartz has grown statically by grain boundary migration. A sinistral sense of shear can be inferred from the curvature or stair stepping of the structure. Luminarias, southern Minas Gerais State, SE Brazil. Width of view 13 mm. CPL.

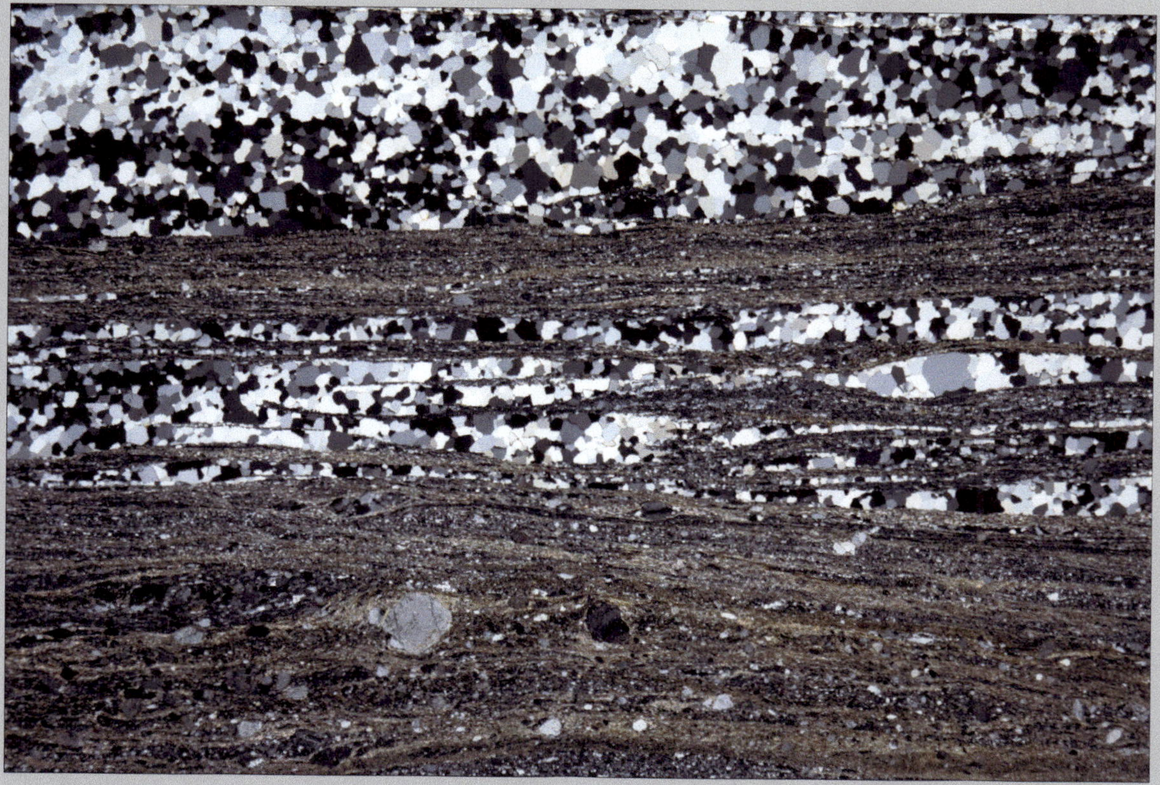

Fig. 5.19 Upper greenschist facies mylonite showing a remarkable contrast between mylonitic and ultramylonitic fabric in the lower part of the photograph. Porphyroclasts of feldspar are surrounded by a mica-rich matrix and quartz veins with a granoblastic fabric show evidence of static grain growth. The sense of shear is dextral but this is not obvious from this photomicrograph. Itumirim, southern Minas Gerais State, SE Brazil. Width of view 20 mm. CPL.

Fig. 5.20 Close-up from the same thin section as Fig. 5.19, showing the polygonal granoblastic fabric in quartz, resulting from static grain growth. Width of view 3 mm. CPL.

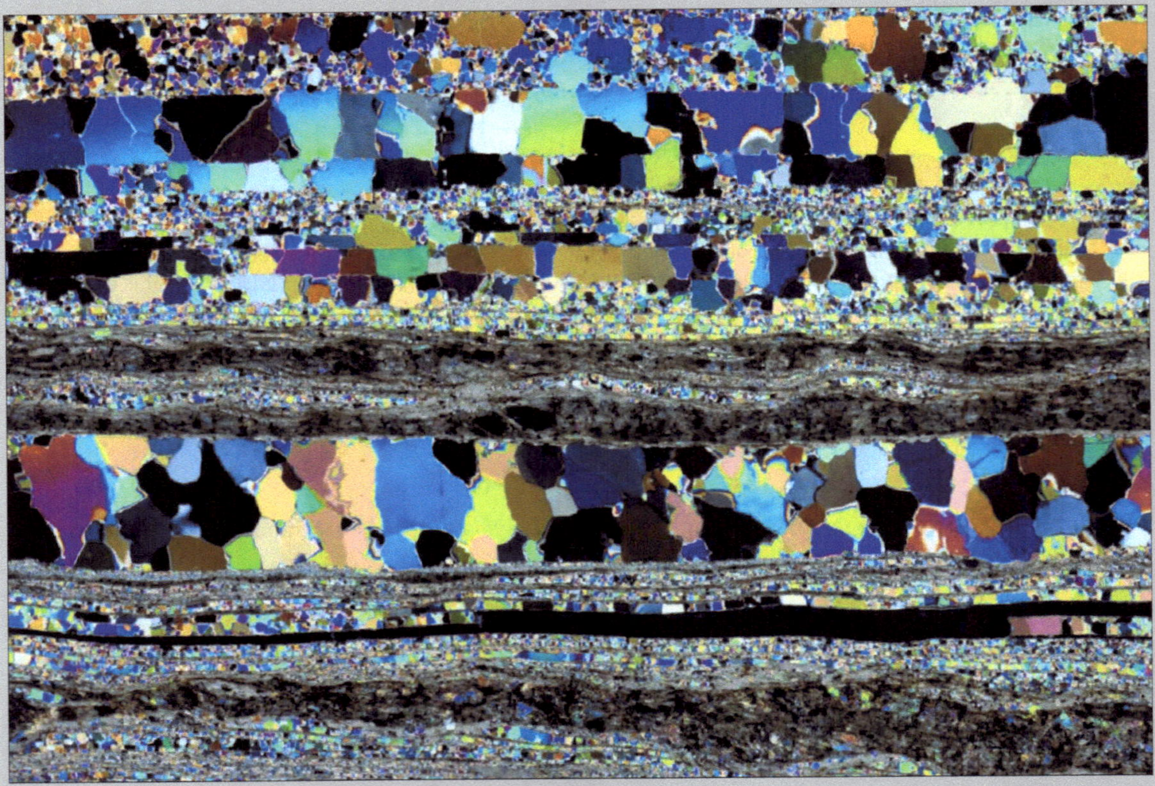

Fig. 5.21 Fine-grained mylonite with recrystallised quartz grains forming a polygonal granoblastic fabric. The strong colour contrast between the individual quartz grains shows that the lattice orientation of adjacent recrystallised quartz grains is quite different. Several asymmetric folds in the fine-grained mylonitic foliation and a poorly developed oblique foliation indicate sinistral sense of shear. Origin unknown.
Width of view 4 mm. CPL with gypsum plate.

Chapter 6 | High-Grade Mylonites

6 High-Grade Mylonites

High-grade mylonites are formed at temperatures above 650 °C. They are relatively uncommon, probably because their conservation is problematic. Most mylonites formed under these conditions would tend to fully recrystallise which destroys and masks the mylonitic structure. Mylonitic features are only preserved if grain growth is somehow inhibited in the rock, e.g. by its polymineralic nature.

Prominent features of high-grade mylonites are monocrystalline ribbons of recrystallised quartz, grown by grain boundary migration to elongated grains that are bordered by recrystallised feldspar. Few feldspar porphyroclasts with core-mantle structures may survive but most feldspar is recrystallised to polygonal crystalloblastic aggregates. Deviation of the foliation around porphyroclasts is usually symmetric and shear sense indicators are rarely visible in thin section.

In some examples recrystallisation of hornblende and orthopyroxene is apparent. Recrystallised grains tend to be large: quartz > 200 micrometer and feldspar > 100 micrometer.

Fig. 6.1 High-grade mylonite, or striped gneiss, derived from pegmatitic granite composed mainly of K-feldspar and quartz. The metamorphic grade during mylonitisation is revealed by the large grain size of recrystallised quartz and by the high percentage and relatively large grain size of recrystallised feldspar. The quartz forms ribbons that probably represent strongly deformed granoblastic grains, now recrystallised by grain boundary migration. The ribbon shape is preserved because along both sides of each ribbon relatively small recrystallised K-feldspar grains impeded growth of quartz. The porphyroclast of K-feldspar, just below the center, shows undulose extinction and subgrain rotation recrystallisation along the rims. Três Rios, Rio de Janeiro State, SE Brazil. Width of view 16 mm. CPL.

Fig. 6.2 High-grade mylonite derived from pegmatitic granite composed mostly of K-feldspar and quartz. Note the two porphyroclasts just below the center and an elongated one with undulose extinction at upper left. Same locality as Fig. 6.1. Width of view 22 mm. CPL.

Fig. 6.3 High-grade mylonite derived from granite. Horizontal quartz ribbons are present in the lower part together with porphyroclasts of K-feldspar. Notice that recrystallisation in the lower porphyroclast advances not only from the rim inwards but also along horizontal high-strain zones, probably by subgrain rotation. K-feldspar in the upper part of the photomicrograph is recrystallised to a larger grain size than in the lower part, probably due to the presence of other minerals. Miguel Pereira, Rio de Janeiro State, SE Brazil. Width of view 16 mm. CPL.

Fig. 6.4 High-grade mylonite with porphyroclast of K-feldspar exhibiting undulose extinction due to crystal-plastic deformation. Progressive recrystallisation is substituting the rim by a mosaic of new grains. Inclusions in the porphyroclast are of granoblastic quartz, with slight undulose extinction, that escaped from the intense deformation being enclosed in the strong porphyroclast. Notice quartz ribbons in the matrix with single elongated quartz grains, due to grain boundary migration. Três Rios, Rio de Janeiro State, SE Brazil. Width of view 22 mm. CPL.

Fig. 6.5 Thin section from the same outcrop as Fig. 6.4. The K-feldspar porphyroclast with strong undulose extinction shows progressive recrystallisation by subgrain rotation along upper right – lower left high-strain bands. Notice the quartz ribbons composed of strain-free elongated quartz in the matrix of relatively fine-grained recrystallised K-feldpar. Width of view 16 mm. CPL.

Fig. 6.6 Detail of Fig. 6.5 showing the transition between the recrystallised and non recrystallised parts within the deformed K-feldspar porphyroclast. New grains have a rather uniform grainsize, reflecting equal differential stress (and temperature), and they appear in areas of high strain due to subgrain rotation. Width of view 4 mm. CPL.

Fig. 6.7 Detail of the same microtructure as shown in Fig. 6.6. The central upper part of the photomicrograph is mainly composed of subgrains that change gradually into new grains down and sidewards.
Width of view 3 mm. CPL.

Fig. 6.8 Medium- to high-grade protomylonite derived from felsic granulite, with plagioclase, orthopyroxene, clinopyroxene, minor biotite and quartz. The matrix is mainly composed of plagioclase with undulose extinction. Pyroxene also exhibits undulose extinction. This rock, with only very minor quartz and biotite, resisted deformation and its classification as a mylonite is questionable (there is no clear distinction between matrix and porphyroclasts). In our opinion the designation as protomylonite or mylonitic granulite seems the most appropriate. Mount West, Western Australia. Width of view 13 mm. CPL.

Fig. 6.9 Detail of Fig. 6.8, showing undulose extinction of plagioclase. The coloured grains are pyroxenes. Width of view 8 mm. CPL.

Fig. 6.10 Detail of Fig. 6.8, with gently folded pyroxene crystal (coloured, above center) and deformed plagioclase with undulose extinction. Note that in this rock no clear distinction can be made between matrix and porphyroclasts. Width of view 2.5 mm. CPL.

Fig. 6.11 Quartz-rich band in high-grade paragneiss composed of quartz, plagioclase, K-feldspar, biotite, garnet and sillimanite. Quartz is recrystallised and grew by high-temperature grain boundary migration to elongate shape with lobular grain boundaries (partly due to pinning). In the white grain at upper left a plagioclase fish can be seen, indicating dextral sense of shear; to the right of the center other fish are present. The rock is not a mylonite in the strict sense, but it must have passed through a mylonitic stage as testified by the asymmetric plagioclase fish. Morro do Cara de Câo, Rio de Janeiro, SE Brazil. Width of view 12 mm. CPL.

Fig. 6.12 High-grade mylonite derived from hornblende granite. Note the hornblende porphyroclasts with strain shadows of opaque material in a matrix of quartz, feldspar and biotite. The porphyroclasts and their strain shadows show a sigmoidal shape with stair stepping to the left, indicating sinistral sense of shear. The matrix is composed of recrystallised feldspar and quartz ribbons indicating growth by high-temperature grain boundary migration. Santo Antônio de Padua, Rio de Janeiro State, SE Brazil. Width of view 16 mm. PPL.

Fig. 6.13 As Fig. 6.12. Width of view 16 mm. CPL.

Fig. 6.14 This is a high-grade gneiss derived from hornblende granite with a mylonitic structure partially masked by recrystallisation. The central green patch is a relic porphyroclast of hornblende now recrystallised to small hornblende and quartz (partially altered in the center). In Fig. 6.15 the completely recrystallised granoblastic texture of the rock is evident, but in Fig. 6.14 the asymmetric shape and the stair stepping to the right of the porphyroclastic aggregate indicate dextral sense of shear. Miguel Pereira, Rio de Janeiro State, SE Brazil. Width of view 15 mm. PPL.

Fig. 6.15 As Fig. 6.14. Width of view 15 mm. CPL.

Fig. 6.16 Mylonitic granulite or high-grade mylonite derived from granulite. Dark green grains are hornblende, light green ones clinopyroxene, pink grains orthopyroxene, and most of the whitish grains are plagioclase. Plagioclase and pyroxene show partial recrystallisation (see details in Figs. 6.18 and 6.19) indicating that the mylonitisation took place under granulite facies or slightly lower conditions. Notice the horizontal shear band (C′) in the center indicating sinistral sense of shear. Três Rios, Rio de Janeiro State, SE Brazil. Width of view 25 mm. PPL.

Fig. 6.17 As Fig. 6.16. Width of view 25 mm. CPL.

Fig. 6.18 Detail of Fig. 6.17, showing a relatively coarse-grained matrix of polygonal recrystallised feldspar, with undulose porphyroclasts of plagioclase, some with tapering twins induced by the deformation. The large dark porphyroclast in the upper part is pyroxene, surrounded by small recrystallised grains. No sense of shear can be deduced from this image. Width of view 3 mm. CPL.

Fig. 6.19 Other detail of Fig. 6.17 with partially recrystallised orthopyroxene (brownish) and polygonal granoblastic plagioclase with undulose extinction. Width of view 1.25 mm. CPL.

Chapter 7 | Protomylonite, Mylonite and Ultramylonite

7 Protomylonite, Mylonite and Ultramylonite

The objective of this chapter is to show how variation of strain intensity can be judged in thin section. Usually this kind of variation can best be observed in low-grade mylonites where the percentage of porphyroclasts decreases progressively with strain intensity. However, the percentage of matrix is highly dependent on mineralogical composition (e.g. quartz and biotite tend to convert to matrix readily). Compositional banding in gneiss can therefore result in mylonitic banding of apparent strain variation, which in fact only reflects variation in composition of the parent rock. Several examples of ultramylonite are derived from quartzitic rocks that tend to form few or no porphyroclasts at all.

The problem with protomylonites, formed at low grade, is that these rocks tend to show evidence of strong deformation by elongated quartz with undulose extinction and broken feldspars, before developing a recognisable mylonitic matrix. In these rocks the percentage of matrix can rarely be used as a practical parameter for classification.

Fig. 7.1 Low-grade mylonite derived from granite. The lower part of the photomicrograph shows a protomylonite with broken feldspars and elongated quartz grains with strong undulose extinction; some mica strings are also present. The percentage of matrix in this part can be estimated as somewhere around 20% but the transition between matrix and porphyroclasts is gradational and not clearly defined. A sharp contact marks the transition to an ultramylonite band, with less than 10% porphyroclasts and the upper part of the photomicrograph shows a mylonite with approximately 60% matrix and 40% porphyroclasts. A dextral sense of shear can be inferred from the slight inclination of the mylonitic foliation (S plane) in the lower protomylonitic part, with respect to the foliation of the ultramylonite (C plane). Saint- Barthélemy Massif, French Pyrenees. Width of view 3 mm. CPL.

Fig. 7.2 Mylonite in between two ultramylonite bands from the same thin section as Fig. 7.1. The abrupt transition indicates strain softening during mylonitisation that tends to channel the deformation along finer-grained parts, which then become ultramylonites. Width of view 5 mm. CPL.

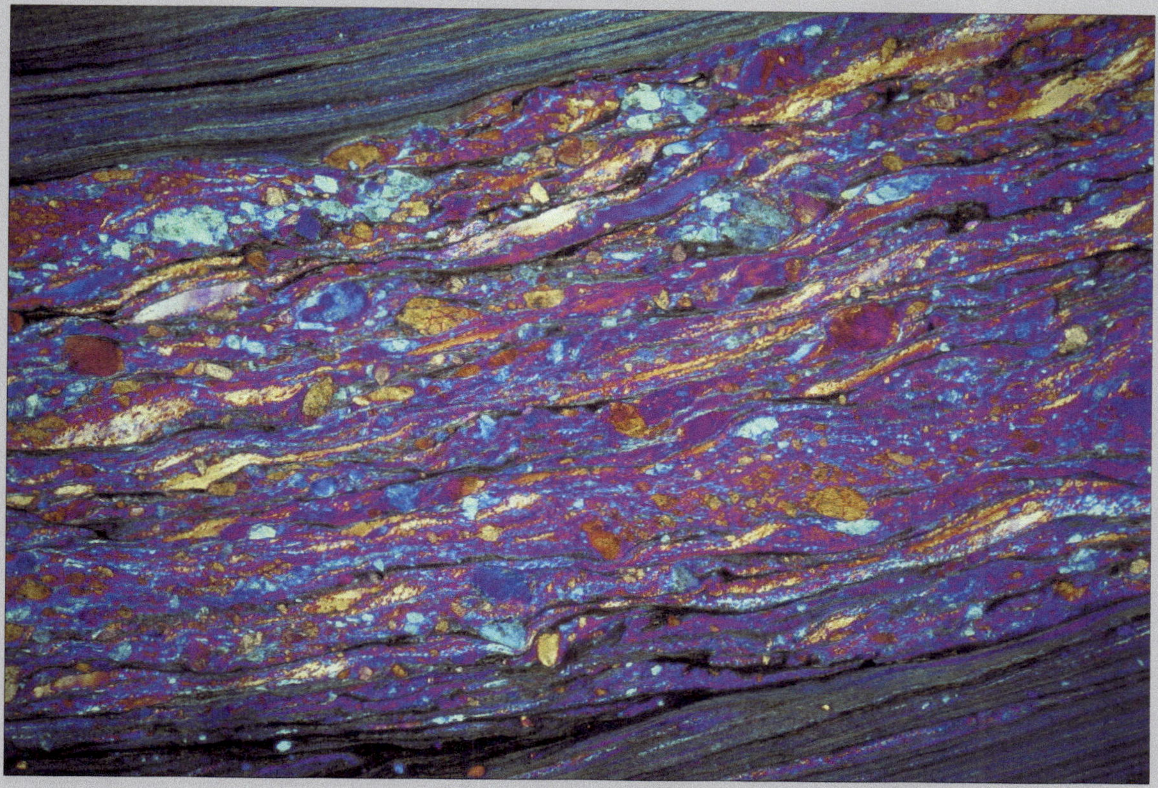

Fig. 7.3 Similar view as in Fig. 7.2, with insertion of gypsum plate to show that elongated quartz lenses, although partially recrystallised, have similar optical orientation. Sense of shear is dextral as can be inferred from stair stepping in the yellow quartz lens left of the center. Width of view 5 mm. CPL with gypsum plate.

Fig. 7.4 Close-up of the same thin section as shown in Figs. 7.1, 7.2 and 7.3, showing deformed and broken feldspar in the lower protomylonitic part, cut by a very fine-grained biotite-rich ultramylonite band, and an upper mylonitic part, with angular feldspar fragments and stretched quartz grains with strong undulose extinction (left of center). The inclined foliation in the upper and lower parts (S planes) combined with the horizontal foliation in the ultramylonite band (C planes) point to dextral sense of shear. Width of view 1.25 mm. CPL.

Fig. 7.5 View with gypsum plate of a contact between ultramylonite and protomylonite (lower part) from the same sample as Fig. 7.1. Feldspar fragments with similar colour are probably derived from a single parent crystal (e.g. the blue fragments below the center), indicating dextral sense of shear (see also the yellow grain at lower right with C´-type shear bands). Width of view 5 mm. CPL with gypsum plate.

Fig. 7.6 Initiation of mylonitisation in granite. The part shown in this view, from the same sample as Fig. 7.1, may be called a protomylonite but it is hard to define what is matrix and what are porphyroclasts. The photomicrograph shows the difference in behaviour between quartz, strongly affected by crystal-plastic deformation, as shown by its undulose extinction (below and left of center), and feldspar that shows much less undulose extinction and tends to be more fractured (upper right and lower left). Biotite (coloured grains) tends to recrystallise to small grains that form strings along the stronger deformed bands (left and right of center). No sense of shear can be determined from this view alone. Width of view 2 mm. CPL.

Fig. 7.7 Mylonite (lower part) and ultramylonite (upper part; same thin section as Fig. 7.1). The porphyroclasts of feldspar are generally rounded but do not show a clear preferred shape orientation. The quartz is stretched out to very elongate lenses with undulose extinction and incipient recrystallisation along the rims (just below and above the center). Some inclined mica fish appear in pink. Sense of shear is defined as dextral by these mica fish and the slight inclination of the quartz lenses. Width of view 3 mm. CPL.

Fig. 7.8 Low-grade ultramylonite derived from a quartzite. Original large quartz grains were strongly deformed by crystal-plastic deformation to zones with undulose extinction. Recrystallisation produced small new grains along the contacts that were also strongly deformed to produce a vague oblique foliation, inclined to the left, indicating dextral sense of shear. Conceição do Rio Verde, southern Minas Gerais State, SE Brazil. Width of view 3 mm. CPL.

Fig. 7.9 Ultramylonite derived from a micaceous quartzite with crossed polarized light and the gypsum plate. Although quartz is extremely deformed (comparable to Fig. 7.8) the colour pattern indicates a strong linking in terms of lattice orientation between old deformed grains and recrystallised grains. The bluish upper left part is probably all derived from a single parent grain and the reddish lower right part idem. No sense of shear can be determined in this view. Conceição do Rio Verde, southern Minas Gerais State, SE Brazil.
Width of view 3 mm. CPL with gypsum plate.

Fig. 7.10 View of the same thin section as Fig. 7.9. Note the difference between the upper left part richer in quartz with either blue or purple colours and the lower right part with more mica and small feldspar grains that shows a more varied colour pattern, due to the smaller grain size of quartz at the onset of deformation.
Width of view 3 mm. CPL with gypsum plate.

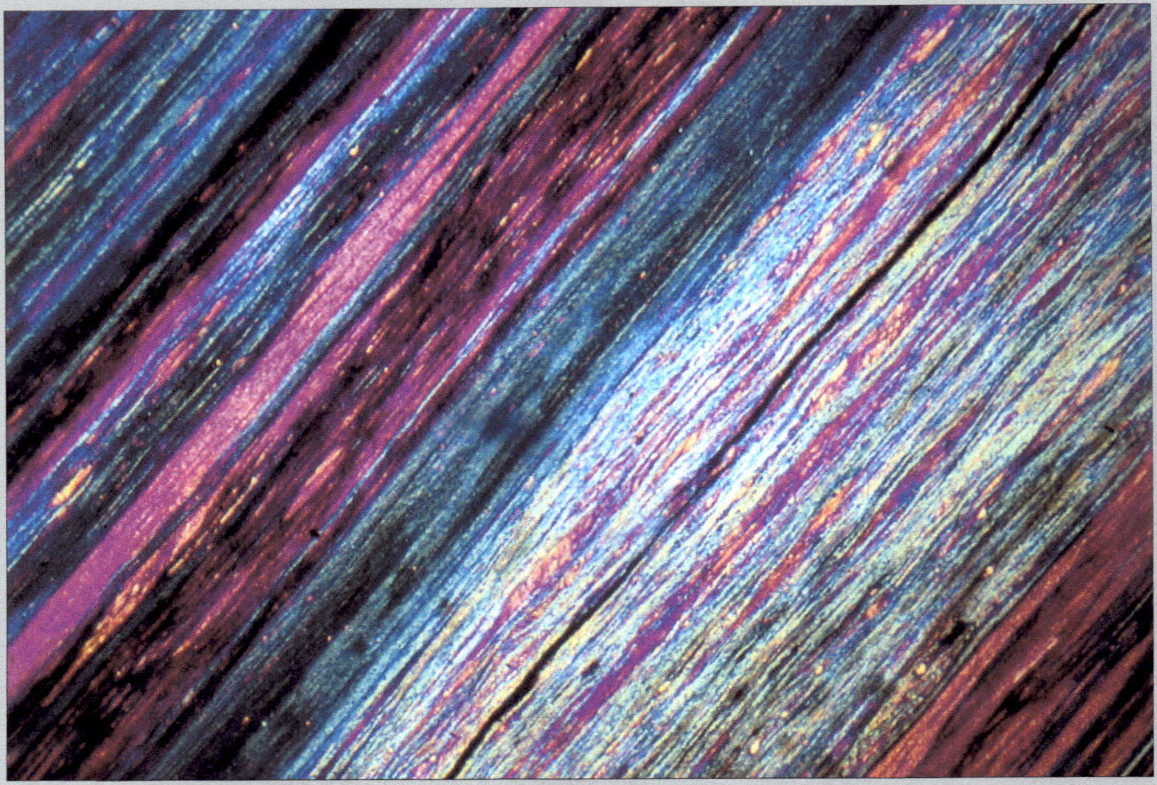

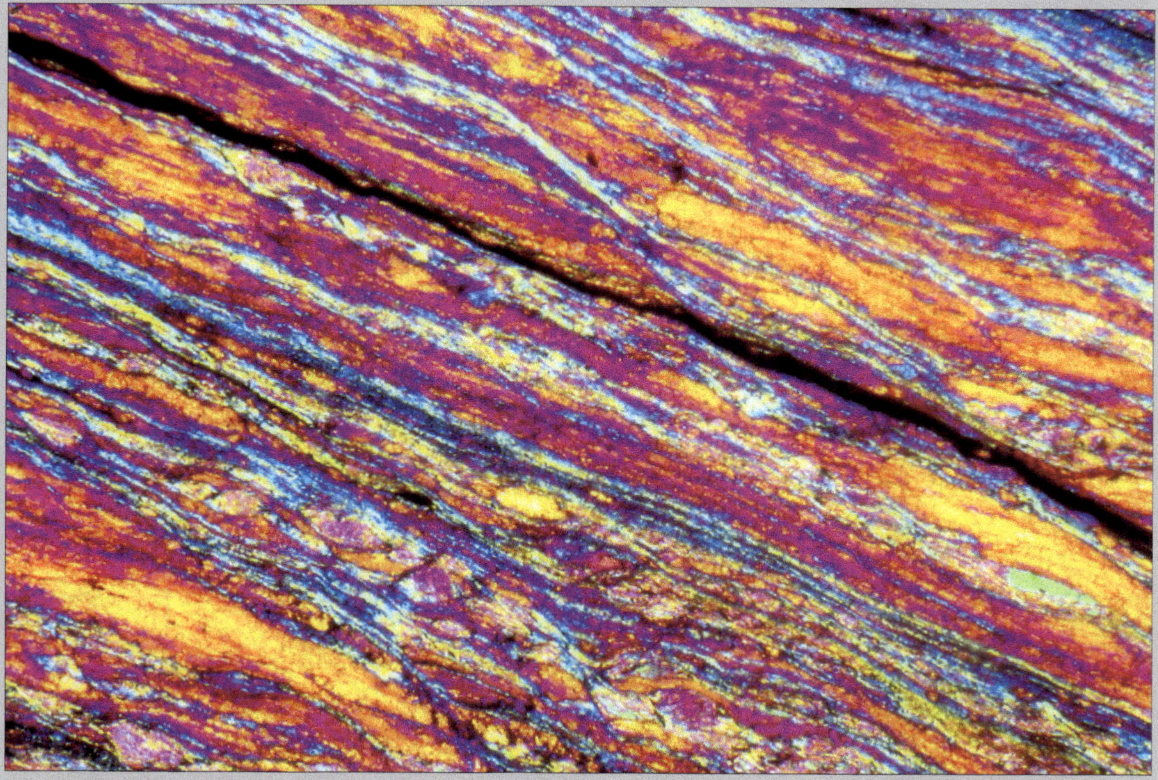

Fig. 7.11 Close-up of the same thin section as Fig. 7.9 to show that, at this magnification, tiny resistant lenses of mica or feldspar become visible, just as micro shear bands (C´, just above the center), indicating dextral sense of shear. Width of view 1 mm. CPL with gypsum plate.

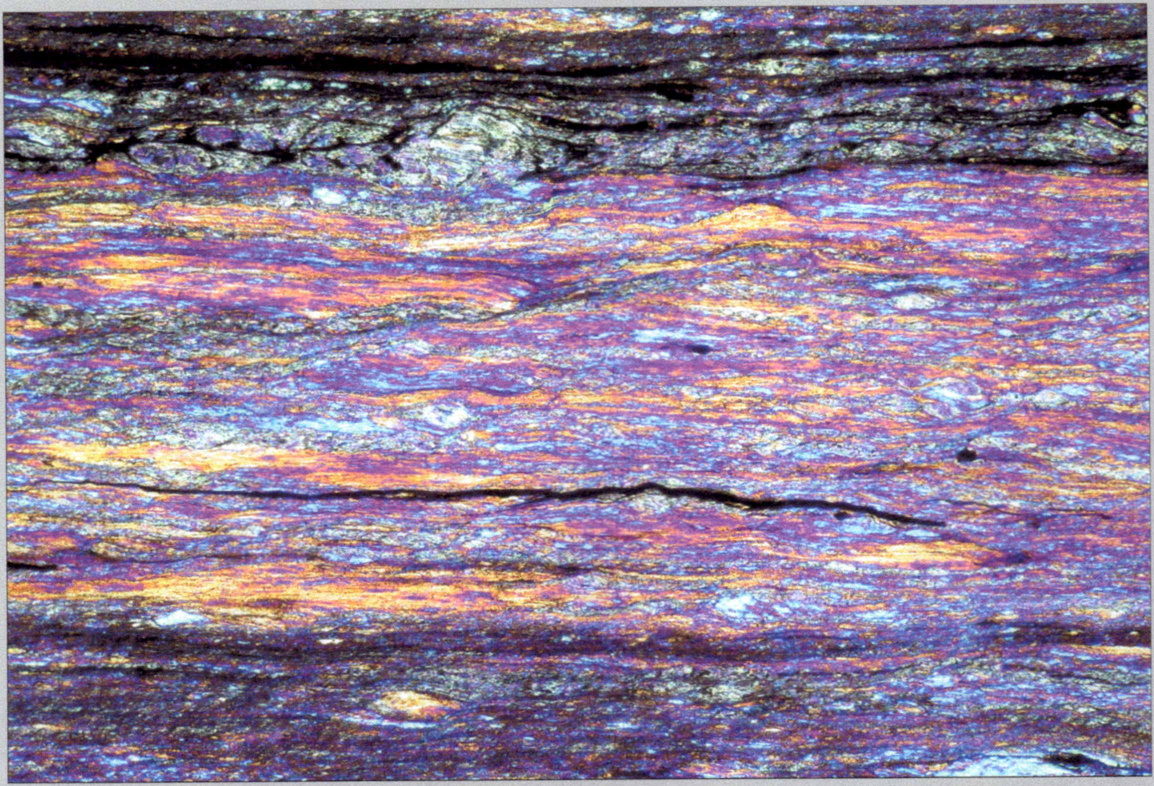

Fig. 7.12 Ultramylonite from the same outcrop as Figs. 7.9 to 7.11, derived from a quartz-feldspathic schist. The shear band (C´) just above and to the left of the center indicates sinistral sense of shear.
Width of view 3 mm. CPL with gypsum plate.

Fig. 7.13 Close-up of Fig. 7.12 to show the inclined shear band more clearly.
Width of view 1.25 mm. CPL with gypsum plate.

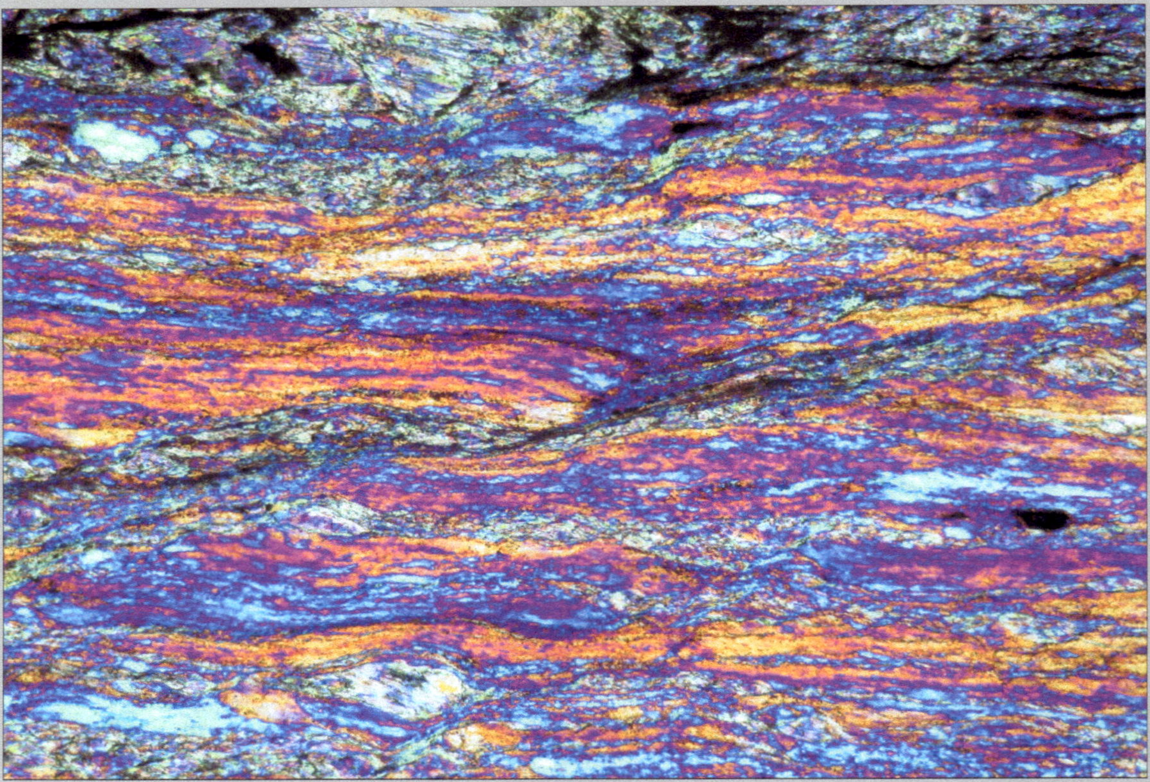

Fig. 7.14 Ultramylonite derived from granite with a few feldspar porphyroclasts and a single muscovite fish (just below the center). It is surprising that muscovite resists as a porphyroclast in this ultramylonite, where all quartz and most feldspar are reduced to very small grain size, mainly by recrystallisation. The inclined shape of the fish indicates dextral sense of shear. Saint-Barthélemy Massif, French Pyrenees. Width of view 3 mm. CPL.

Fig. 7.15 Same view as Fig. 7.14, with gypsum plate to show that the small recrystallised quartz grains in the purple bands retain a very similar lattice orientation. Notice a small feldspar porphyroclast just below the mica fish, also indicating dextral sense of shear by stair stepping. CPL with gypsum plate.

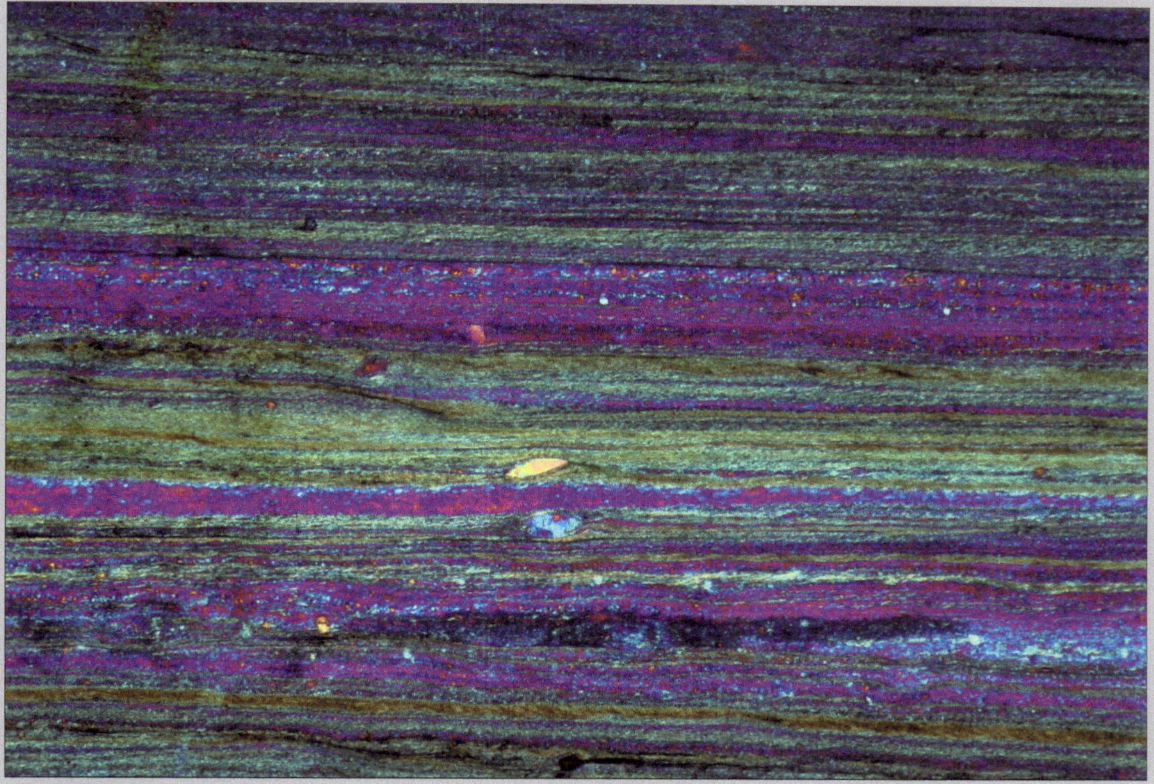

Fig. 7.16 Most parts of this photomicrograph show an ultramylonite derived from a paragneiss, surrounding some lenses of feldspar porphyroclasts. The strong deviation of the mylonitic foliation around these lenses indicates a strong strain gradient between the resistant lenses and the ultramylonite matrix. Sense of shear is dextral, but no reliable shear sense indicators appear in this photomicrograph. Saint-Barthélemy Massif, French Pyrenees. Width of view 12 mm. PPL.

Fig. 7.17 As Fig. 7.16. Width of view 12 mm. CPL.

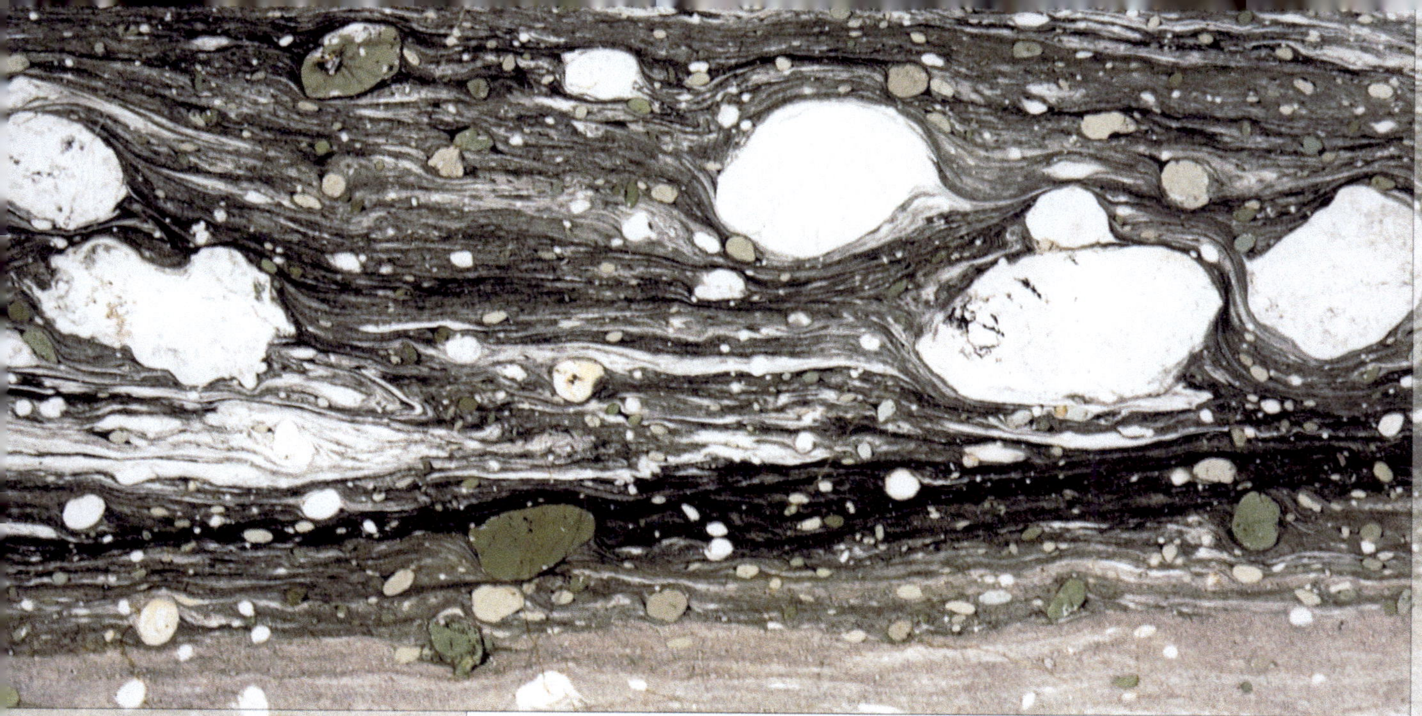

Chapter 8 | Mylonites Derived From Parent Rocks Other Than Granites and Gneisses

8 Mylonites Derived From Parent Rocks Other Than Granites and Gneisses

Most mylonites shown in this atlas are derived from granites and gneisses. This is not a coincidence; the mineralogy of these rocks favours the formation of mylonites because of the contrasting behaviour of quartz and biotite on the one hand (forming matrix) and feldspar and muscovite on the other hand (forming porphyroclasts). Another group of rocks that readily forms mylonites are impure quartzites in which resistant minerals tend to form fish-like structures, again, by strong contrast in rheological behaviour.

In this chapter a number of examples of mylonites derived from other parent rocks is shown.

Carbonate rocks, including marbles, can develop shear zones without much difficulty, but these are not frequently preserved because calcite recrystallises easily even at very low temperatures. Calcsilicate rocks with minarals such as garnet and diopside can form mylonites with these minerals as porphyroclasts in a quartz or carbonate-rich matrix.

Amphibolites, diorites and other amphibole-plagioclase rich rocks are more resistant to mylonitisation than granites and gneisses because of their low content of quartz and biotite. Hornblende is more resistant to deformation than plagioclase and recrystallises only at relatively high temperature; therefore, it usually forms naked clasts in a fine-grained matrix. In some examples, the higher temperature hornblende is surrounded by a lower-grade amphibole, showing evidence of retrograde reactions influenced by the mylonitization.

Mylonitised eclogites with symplectitic clinopyroxene produce interesting mylonites, with garnet porphyroclasts embedded in a fine-grained matrix composed of retrograde products of the destabilisation of clinopyroxene.

Peridotites can produce beautiful mylonites (Fig. 8.1). As in granites, peridotites can form "low-temperature" fabrics with pyroxene porphyroclasts in a fine-grained recrystallised matrix of olivine, and "high-temperature" fabrics with a ribbon like structure and larger grains in the matrix. Although the microstructure forms in the same way as in low- and high-temperature granite mylonites, the actual temperatures are much higher in peridotites.

Serpentinites are easy to deform and tend to develop conspicuous shear bands (Figs. 8.2-8.4).

Fig. 8.1 Mylonitised lherzolite with porphyroclasts of olivine (purple and other bright colours), orthopyroxene (grey), clinopyroxene (yellow-green, left of center) and smaller spinel (black). The matrix is an ultrafine grained mixture of olivine, orthopyroxene, clinopyroxene, plagioclase and spinel. Dextral sense of shear is indicated by the inclined fish shape of several porphyroclasts and by asymmetric folds in the matrix, verging to the right. The shear zone was formed under granulite facies conditions during exhumation of the upper mantle in a zone of transcurrent rifting during the Cretaceous (Newman et al. 1999). Note that the structure is comparable to low-grade mylonites from the Earth crust. Turon de las Técouère peridotite body, western Spanish Pyrenees. Width of view 15 mm. CPL (photomicrograph courtesy Martyn Drury).

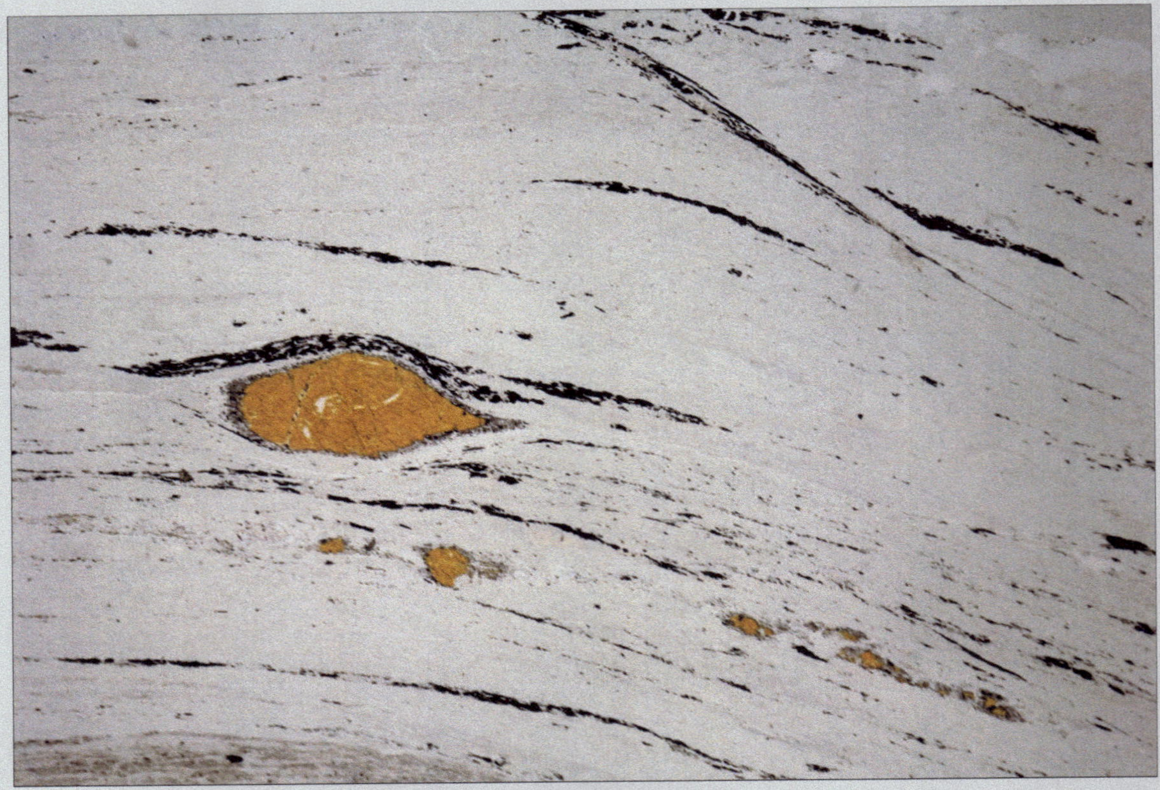

Fig. 8.2 Mylonitic serpentinite with a large porphyroclast of titanoclinohumite. At upper right a C′ shear band, inclined to the right, indicates dextral sense of shear; some carbonate is present along this shear band. Note that the deviation around the porphyroclast is symmetric and does not reveal shear sense. Zermatt, Swiss Alps. Width of view 16 mm. CPL.

Fig. 8.3 As Fig. 8.2. Width of view 16 mm. PPL.

Fig. 8.4 Detail of Fig. 8.3, showing reduction of grain size in the shear band. Width of view 10mm. CPL.

Fig. 8.5 Close-up of Fig. 8.4 showing elongated serpentine crystals, defining an oblique foliation in the shear band, consistent with dextral sense of shear. Width of view 3 mm. CPL.

Fig. 8.6 Serpentinite with carbonate veins that show dextral shear. The rock is not a mylonite but the sheared vein shows how carbonate behaves in shear zones mainly by crystal-plastic deformation. Leiden collection. Width of view 18 mm. CPL.

Fig. 8.7 Sheared carbonate vein showing dextral sense of shear. Notice the transition between the granoblastic fabric at lower left to a strong mylonitic foliation defined by highly elongated grains at center right. The change in orientation of this foliation, from inclined to the left to horizontal, reflects the change in orientation of the XY plane of strain, from low to high strain. Leiden collection. Width of view 12 mm. CPL.

Fig. 8.8 Close-up of Fig. 8.7 showing the variation through a horizontal dextral shear band from low strain along the bottom, to high strain in the central part and to low strain again in the upper part. Note the change in inclination of the foliation, grain size and grain shape across the shear band. Width of view 5 mm. CPL.

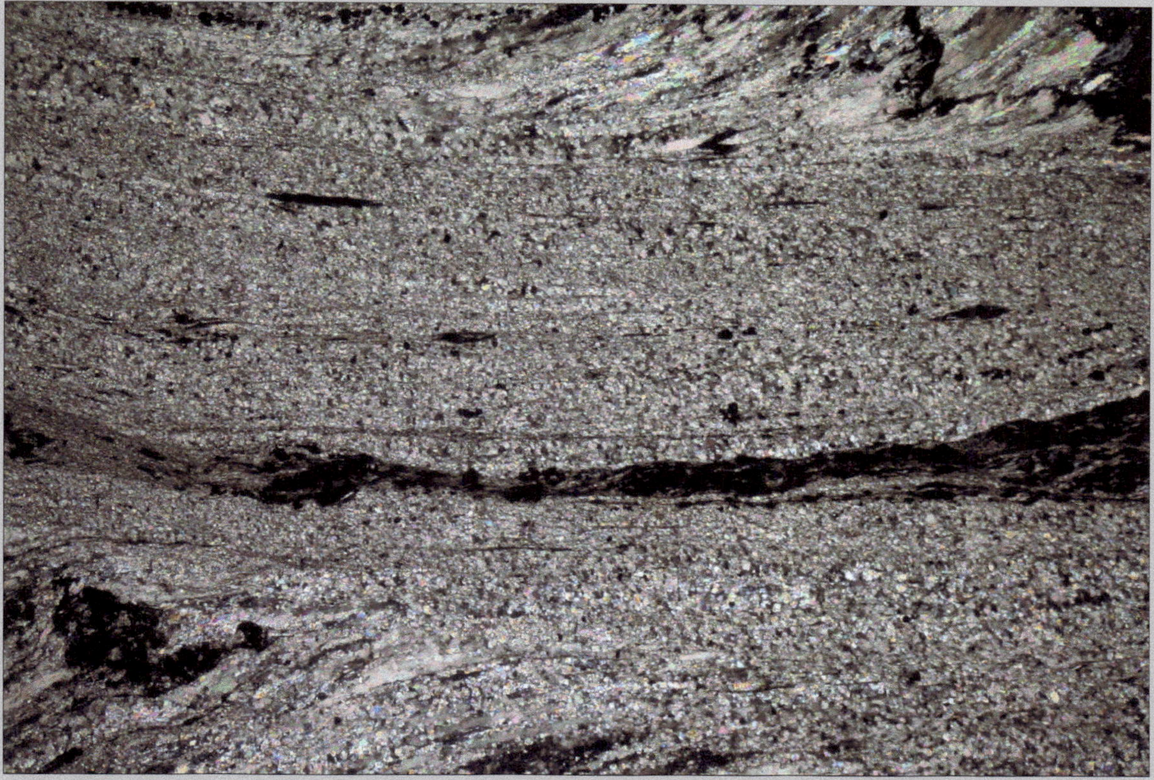

Fig. 8.9 Example of a low-grade mylonitised amphibolite containing naked porphyroclasts of hornblende (green), plagioclase (white) and an opaque mineral (black). The matrix is composed of fine-grained amphibole, biotite, chlorite and some quartz. Note that the dark bands contain more hornblende porphyroclasts and the lighter bands more plagioclase, probably reflecting original compositional banding in the amphibolite. Shear sense is not obvious from this photomicrograph. Note two sharp contacts in the upper part of the photomicrograph where apparently faulting occurred involving brittle deformation. Marsfjällen, Västerbotten, Sweden. Width of view 15 mm. PPL.

Fig. 8.10 As Fig. 8.9. Width of view 15 mm. CPL.

Fig. 8.11 Detail of Fig. 8.9 showing a hornblende porphyroclast cut by a fault with opaque material concentrated along the fault plane. Width of view 2 mm. PPL.

Fig. 8.12 Detail of Fig. 8.9 with hornblende porphyroclast showing a rim of different colour (bright green), reflecting different composition, probably related to a retrograde reaction during mylonitisation. Width of view 2 mm. PPL.

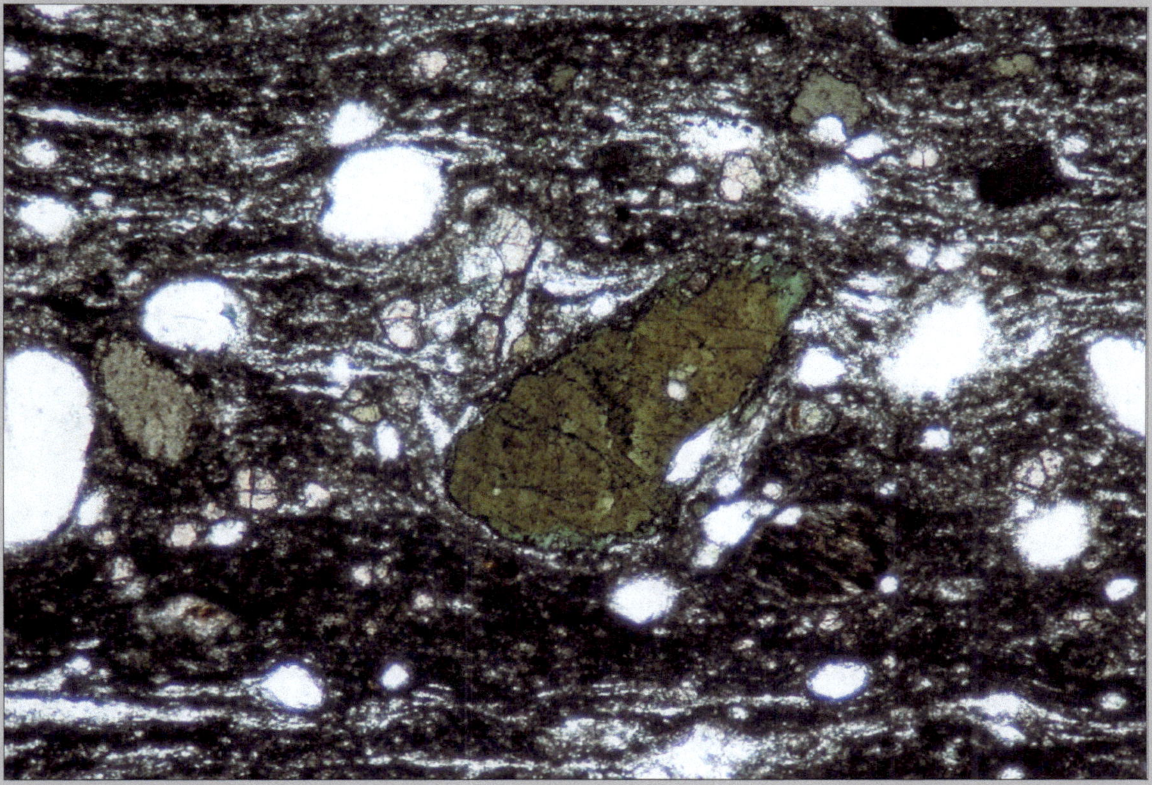

Fig. 8.13 Amphibolite mylonitised under low to medium-grade metamorphic conditions. The porphyroclasts are hornblende, plagioclase and some pyroxenes in the lower dark layer. The dark and light coloured bands reflect compositional banding in the parent amphibolite. Notice that hornblende clasts are naked but plagioclase porphyroclasts have wings. A sinistral sense of shear can be inferred from small asymmetric folds especially in the upper white layer. Western Australia. Width of view 16 mm. PPL.

Fig. 8.14 As Fig. 8.13. Width of view 16 mm. CPL.

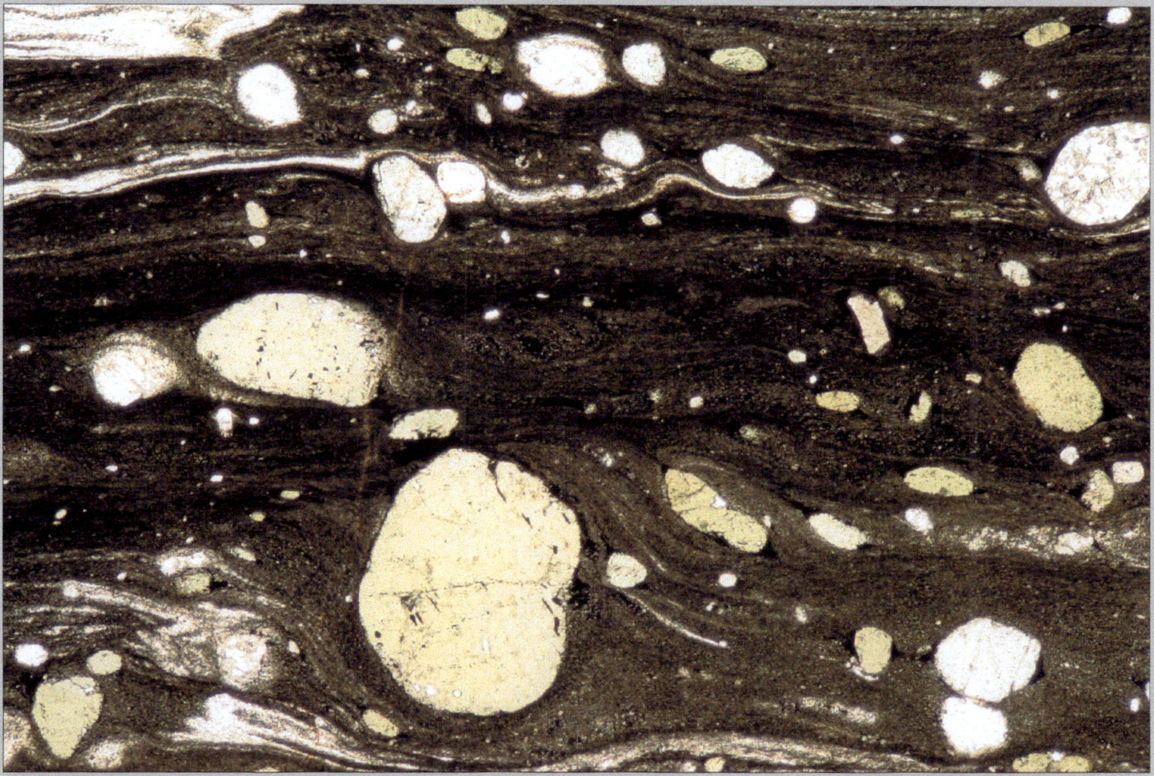

Fig. 8.15 Detail of Fig. 8.13 to show naked clasts of hornblende surrounded by a dark matrix, probably also composed of very fine-grained amphibole. Asymmetric folds indicate sinistral sense of shear. Width of view 2.5 mm. PPL.

Fig. 8.16 Calcsilicate rock composed of quartz, carbonate, plagioclase, biotite, hornblende and chlorite. Sinistral C′ shear bands are developed from upper right to lower left. The rock is not really a mylonite. Chlorite is concentrated in the shear bands (Figs. 8.18 and 8.19) suggesting retrograde (probably greenschist facies) conditions during the formation of the shear bands. New Zealand. Width of view 24 mm. PPL.

Fig. 8.17 As Fig. 8.16. Width of view 24 mm. CPL.

Fig. 8.18 Detail of Fig. 8.16 to show a disrupted hornblende crystal offset in a sinistral sense along a C′ type shear band (upper right to lower left). Width of view 5 mm. PPL.

Fig. 8.19 As Fig. 8.18. Width of view 5 mm. CPL.

Fig. 8.20 Amphibolite, mylonitised at low to medium-grade conditions. Porphyroclasts are of hornblende and plagioclase. Note that hornblende forms naked clasts whereas plagioclase has minor wings. In Fig. 8.21 partial recrystallisation around cores with undulose extinction produced core-mantle structures. Asymmetric folds (left of center) and a delta clast (above center) indicate sinistral sense of shear. Western Australia.
Width of view 16 mm PPL.

Fig. 8.21 As Fig. 8.20. Width of view 16 mm. CPL.

Fig. 8.22 Detail of Fig. 8.20 showing a strong contrast between the very fine-grained dark matrix and apparently strain free hornblende porphyroclasts with some preserved biotite inclusions. Sinistral shear sense can be inferred from stair stepping around the large hornblende porphyroclast above the center. Width of view 3 mm. PPL.

Fig. 8.23 The lower part of this photomicrograph shows an amphibolite, probably derived from diorite with igneous texture partially preserved (no schistosity is apparent). The upper part is affected by a shear zone that reduced the grain size considerably and also changed the mineralogical and chemical composition: the lower band of the shear zone is enriched in hornblende and the upper band is enriched in biotite. A dextral sense of shear can be inferred from the oblique orientation of the foliation in the shear zone. São João del Rei, southern Minas Gerais State, SE Brazil. Width of view 20 mm. PPL.

Fig. 8.24 As Fig. 8.23. Width of view 20 mm. CPL.

Fig. 8.25 C´ shear band in an amphibolite. The green minerals are hornblende and the white ones plagioclase. The inclined position of the shear band (upper right lower left) indicates sinistral sense of shear, confirmed by C/S fabric in the upper left corner with horizontal C planes and S planes inclined to the right. The shear band is mainly composed of actinolite and epidote suggesting greenschist facies conditions during shear movements. São João del Rei, southern Minas Gerais State, SE Brazil. Width of view 40 mm. PPL.

Fig. 8.26 As Fig. 8.25. Width of view 40 mm. CPL.

Fig. 8.27 Mylonitised greenschist with porphyroclasts of epidote and smaller ones of albite embedded in a matrix rich in chlorite. No shear sense is apparent. Equador. Width of view 16 mm. PPL.

Fig. 8.28 As Fig. 8.27. Width of view 16 mm. CPL.

Fig. 8.29 The lower part of this photomicrograph shows a mylonite derived from an amphibolite with porphyroclasts almost exclusively composed of hornblende. The upper part contains porphyroclasts of hornblende, plagioclase and garnet, embedded in a biotite-rich matrix and is probably derived from a hornblende bearing garnet-biotite schist. Shear sense in the mylonite cannot be inferred with certainty, but a late brittle fault (upper left to lower right) shows dextral offset. Scotland. Width of view 16 mm. PPL.

Fig. 8.30 As Fig. 8.29. Width of view 16 mm. CPL.

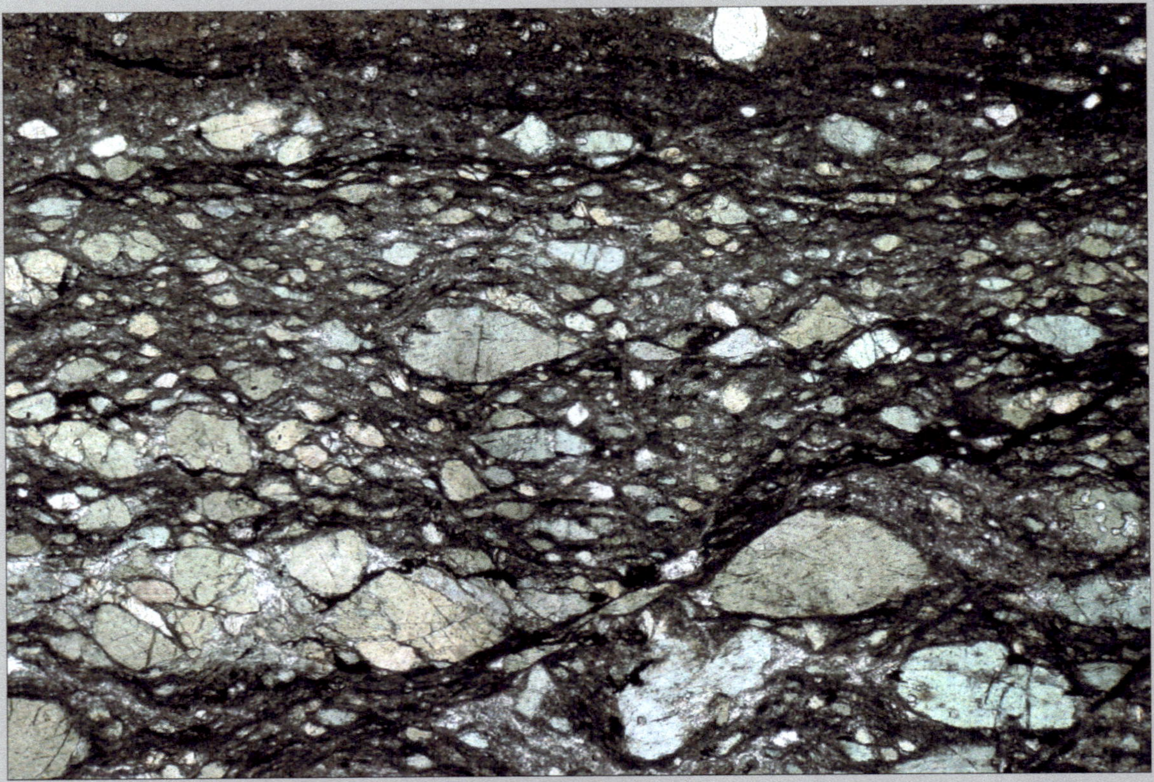

Fig. 8.31 Detail of Fig. 8.29, showing hornblende porphyroclasts embedded in a dark matrix, probably also composed of fine-grained hornblende. A fault at lower right, trending from upper right to lower left, appears to show sinistral displacement. If interpreted as a C´-type shear band it would indicate sinistral sense of shear in the mylonite. Width of view 3 mm. PPL.

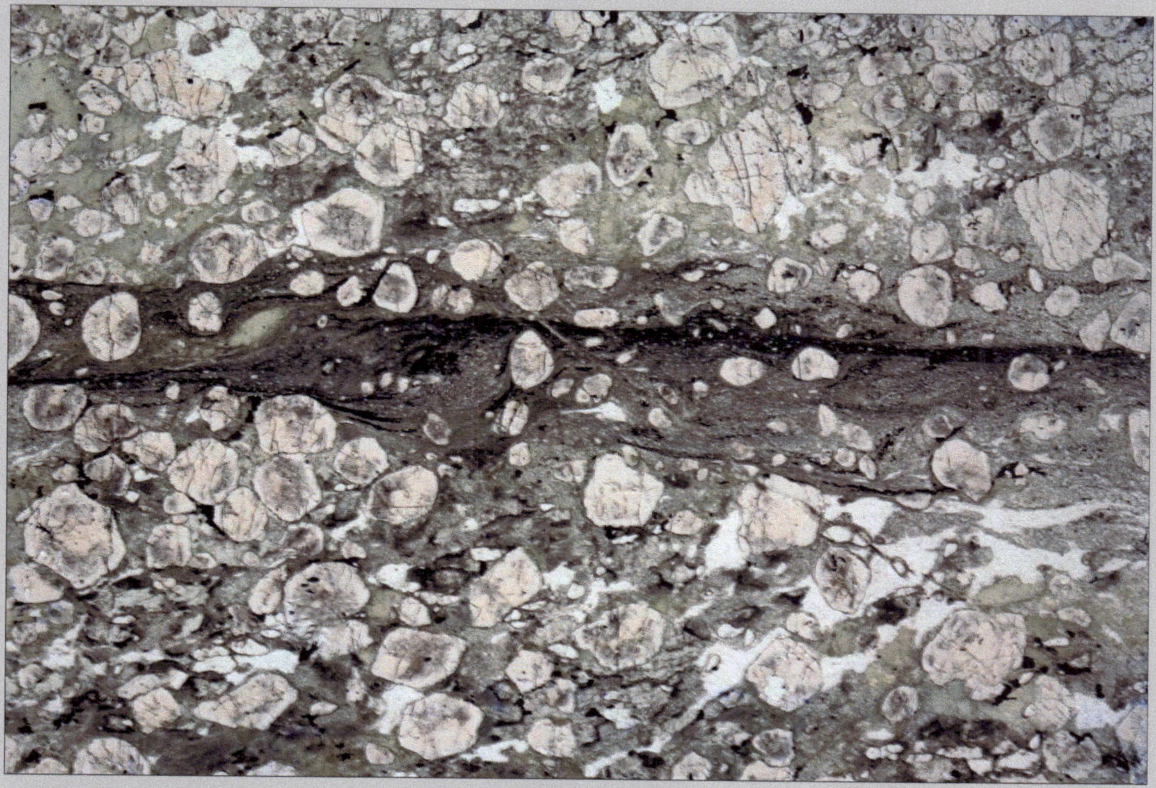

Fig. 8.32 Eclogite composed essentially of garnet and clinopyroxene, well preserved in the upper and lower part. A minor shear zone transects the field of view horizontally. Symplectitic omphacite is readily affected by the shear zone, whereas garnet resists and forms rounded porphyroclasts. Sense of shear is dextral as can be inferred from an oblique lens of omphacite in the left hand part of the shear zone. Cabo Ortegal, Galícia, NW Spain. Width of view 14 mm. PPL.

Fig. 8.33 As Fig. 8.32. Width of view 14 mm. CPL.

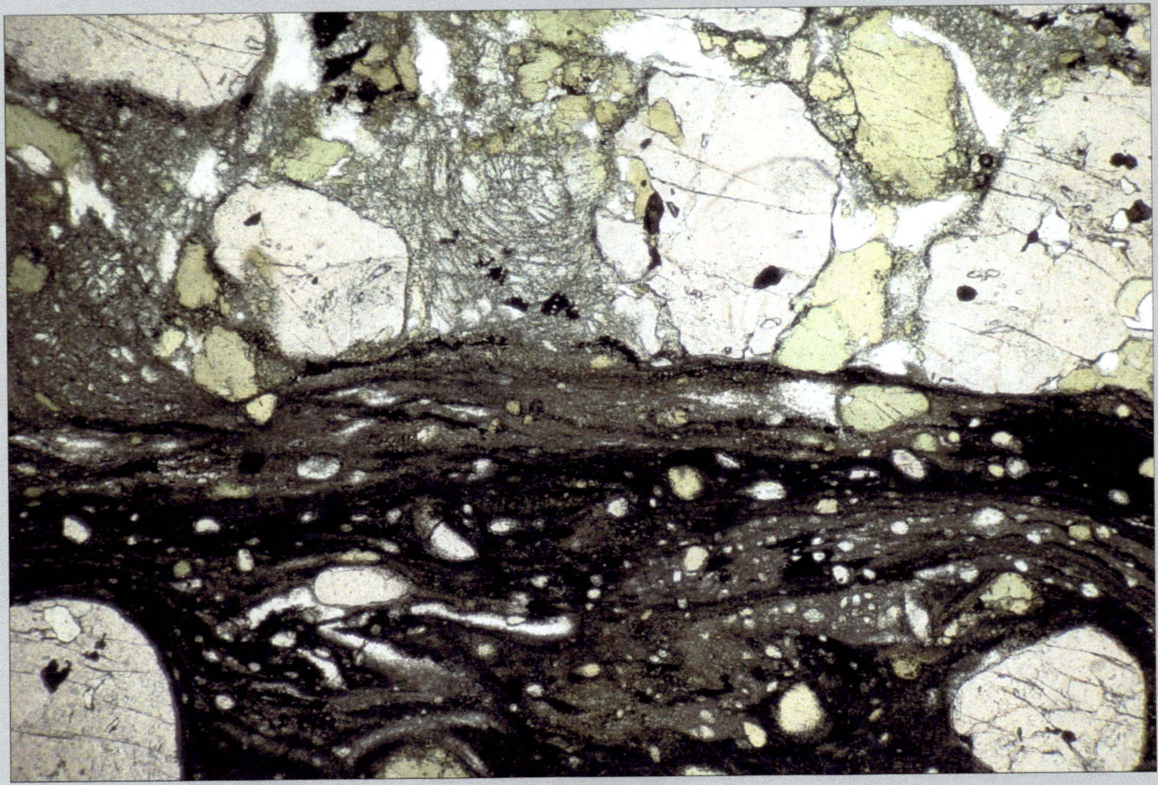

Fig. 8.34 Detail of other shear zones from the same eclogite as Figs 8.32 and 8.33. Symplectitic omphacite appears in the upper part of the photomicrograph, next to garnet and light green amphibole. The dark grains are rutile. A sinistral shear sense can be inferred from the inclined mylonitic foliation close to the transition from shear zone to host rock. Width of view 3 mm. PPL.

Fig. 8.35 As Fig. 8.34. Width of view 3 mm. PPL.

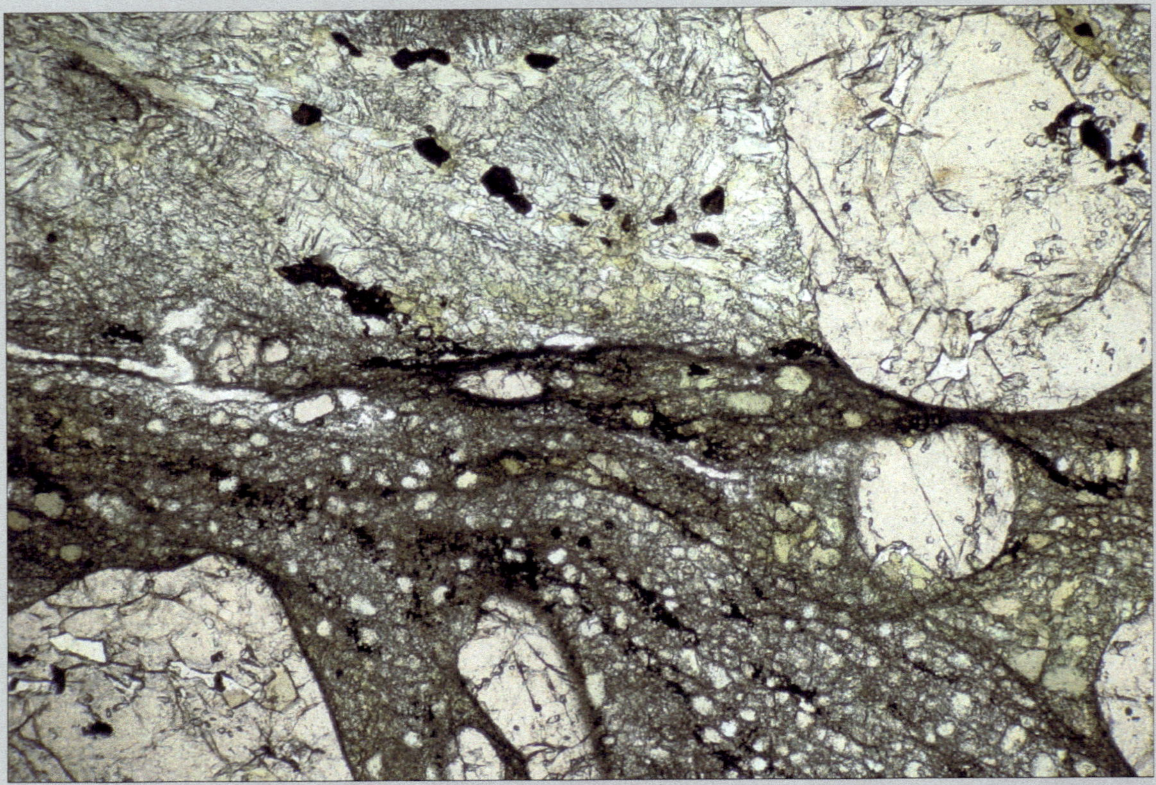

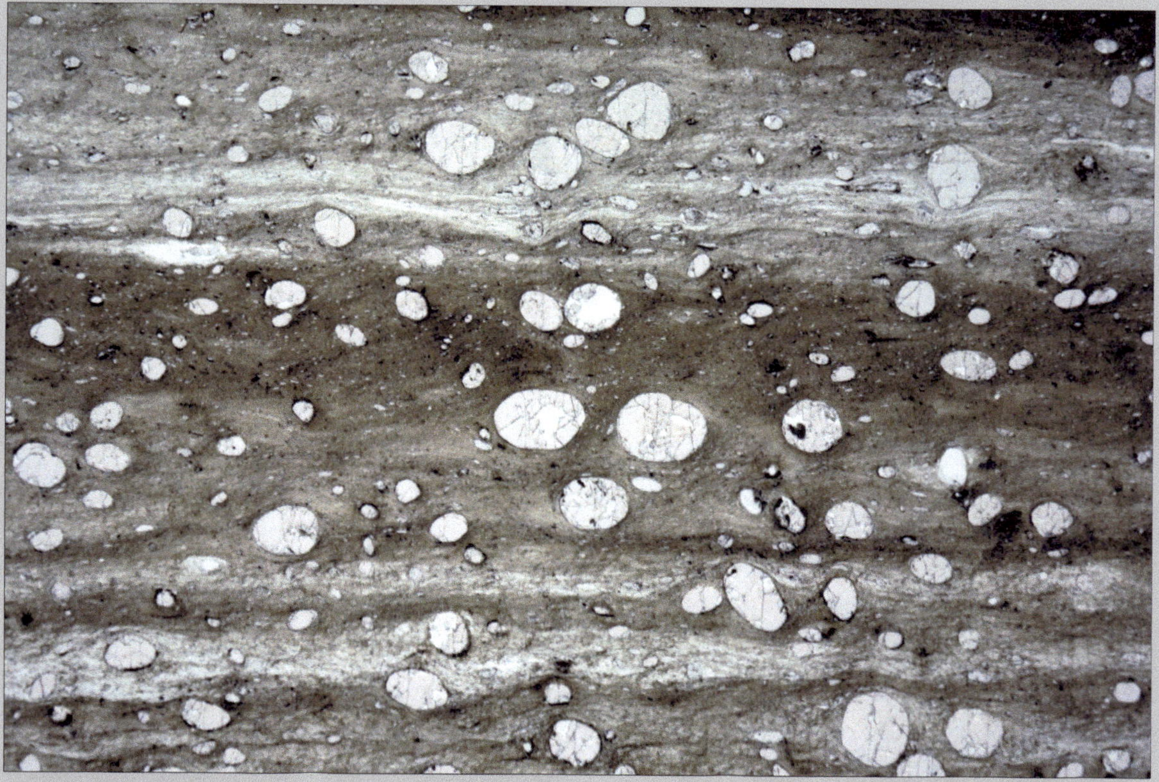

Fig. 8.36 Low-grade mylonite derived from a rock mainly composed of biotite, garnet and kyanite, with minor plagioclase, K-feldspar and quartz. Most porphyroclasts are garnet but several smaller ones are kyanite, showing strong undulose extinction. Note the rounded shape of the garnet porphyroclasts apparently caused by mechanical erosion due to friction against the matrix. No shear sense is apparent. Marsfjällen, Västerbotten, Sweden. Width of view 12 mm. PPL.

Fig. 8.37 As Fig. 8.36. Width of view 12 mm. CPL.

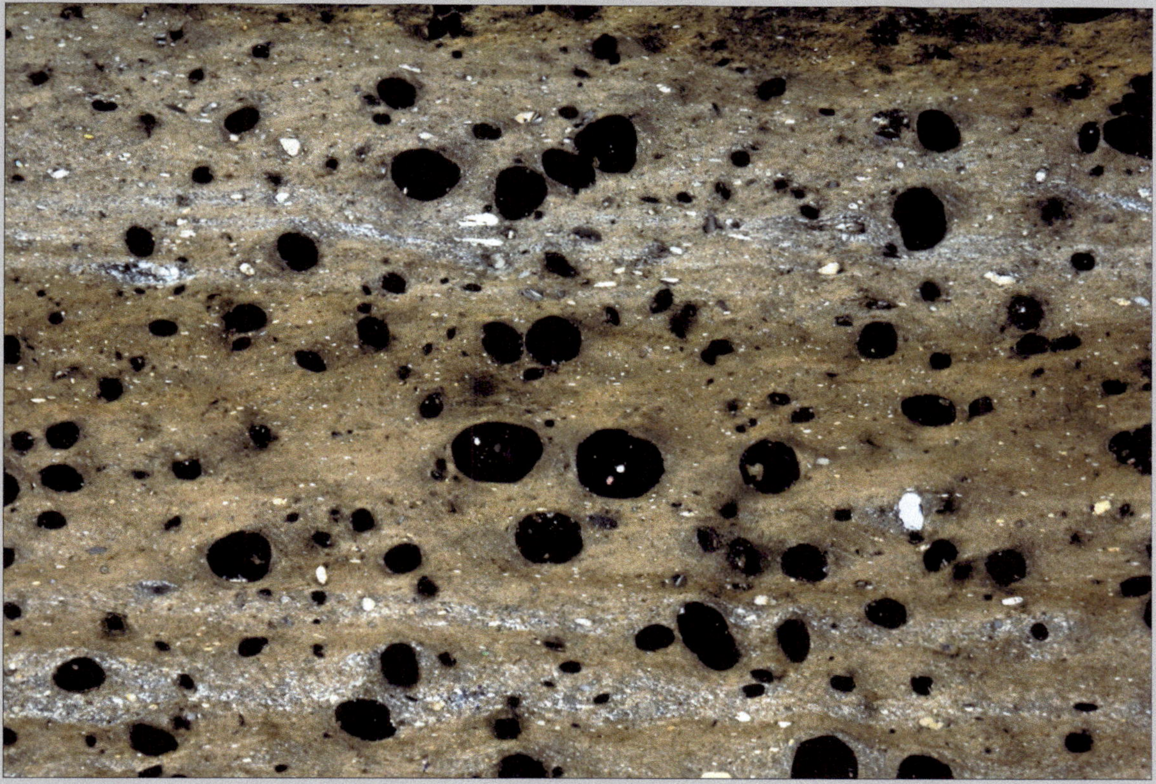

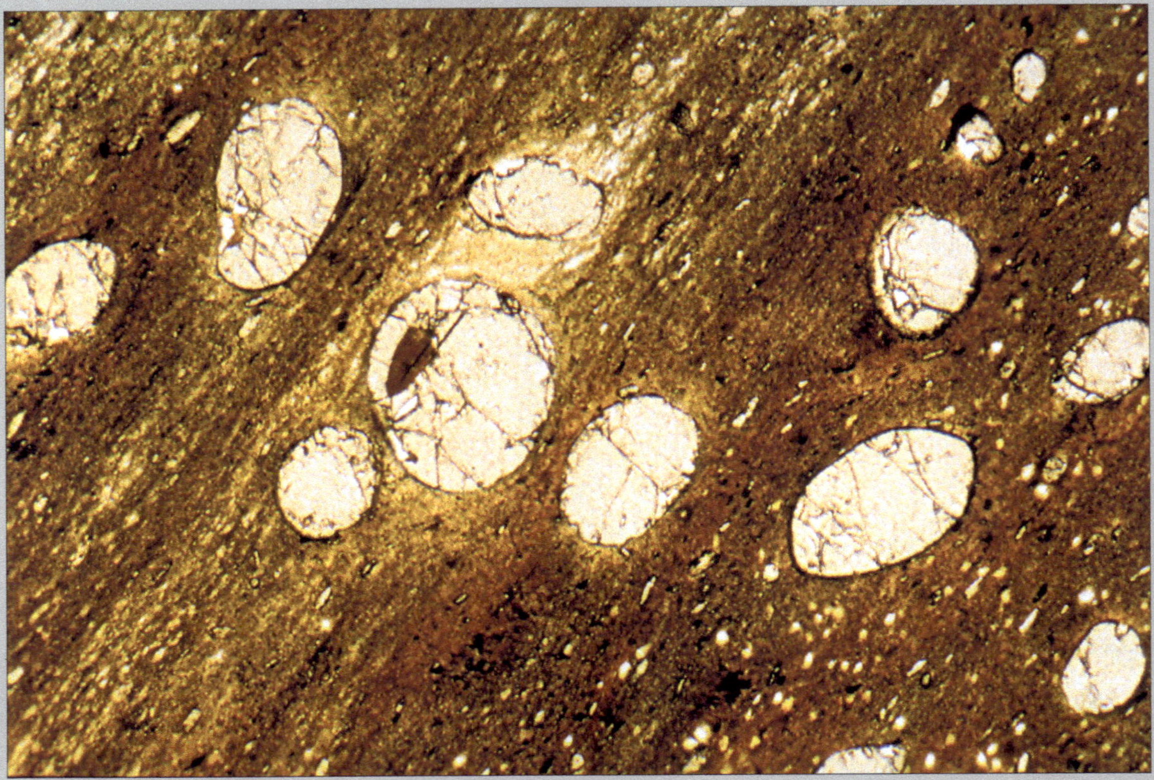

Fig. 8.38 Detail of Fig. 8.36. Note the rounded shape of the garnets and a biotite inclusion in one of the garnets, preserved from the grain size reduction in the matrix. Width of view 5 mm. PPL.

Fig. 8.39 Mylonitised diabase with porphyroclasts of plagioclase and clinopyroxene embedded in a dark matrix. The elongated shape of the plagioclase porphyroclasts is probably inherited from elongated plagioclase laths in a fine-grained matrix of the igneous parent rock. Western Australia. Width of view 12 mm. CPL.

Fig. 8.40 Detail of Fig. 8.39, showing elongated plagioclase laths, rounded at the tips by the mylonitisation. The coloured porphyroclasts are clinopyroxene, also rounded by the mylonitisation. No sense of shear is apparent in this view. Width of view 4 mm. CPL.

Fig. 8.41 Diabase with a large plagioclase phenocryst. A brittle fracture zone filled with small plagioclase fragments cuts the rock in a vertical orientation to the right of the center. Origin unknown. Width of view 3 mm CPL.

Fig. 8.42 Detail of Fig. 8.41, showing the irregularity of the fracture zone filled with very tiny fragments of plagioclase. Width of view 1 mm CPL.

Chapter 9 | Shear Sense Indicators

9 Shear Sense Indicators

Many geologists study mylonites with the exclusive aim to determine the sense of shear. Obviously this is an important aspect, but it is important to study shear zones first, before shear sense determination is attempted. In order to deduce the correct sense of shear we recommend the following procedure (Fig. 9.1)

- Cut the thin section perpendicular to the mylonitic foliation and parallel to the stretching or aggregate lineation in an oriented sample.
- Check the position of the thin section with respect to the sample.
- If the shear zone is subhorizontal, do not use nomenclature such as sinistral or dextral sense of shear, since this depends on the position of the observer and may result in ambiguous descriptions. Use instead nomenclature like "top to the north", or "top to the southeast".
- Do not "force" a sense of shear from rocks that do not show this clearly. In general, the recognition of various different shear sense indicators, indicating the same sense of shear increases reliability. If only one shear sense indicator is found some caution is needed. Shear sense indicators occur in many mylonites but definitely not in all. We recommend to be cautious and, if in doubt, to admit this rather then to guess. It seems to be fashionable to extract shear sense, not only from mylonites but also from many non-mylonitic rocks. In some cases this may be correct, but in many cases it is pointless.

The most important shear sense indicators are summarised in Figure 9.1 (corresponding to Fig. 5.10 of Passchier & Trouw 2005). Most of these are treated in the sections below with the exception of "marker and foliation deflection" and "stair stepping". The offset of marker horizons can rarely be observed in a thin section, although deflections are addressed for C/S and C′ shear-band structures. Stair stepping is a very usefull shear sense indicator, but it occurs invariably in combination with one of the structures discussed below (especially sigma, delta and fish structures) and is therefore not treated as a separate structure.

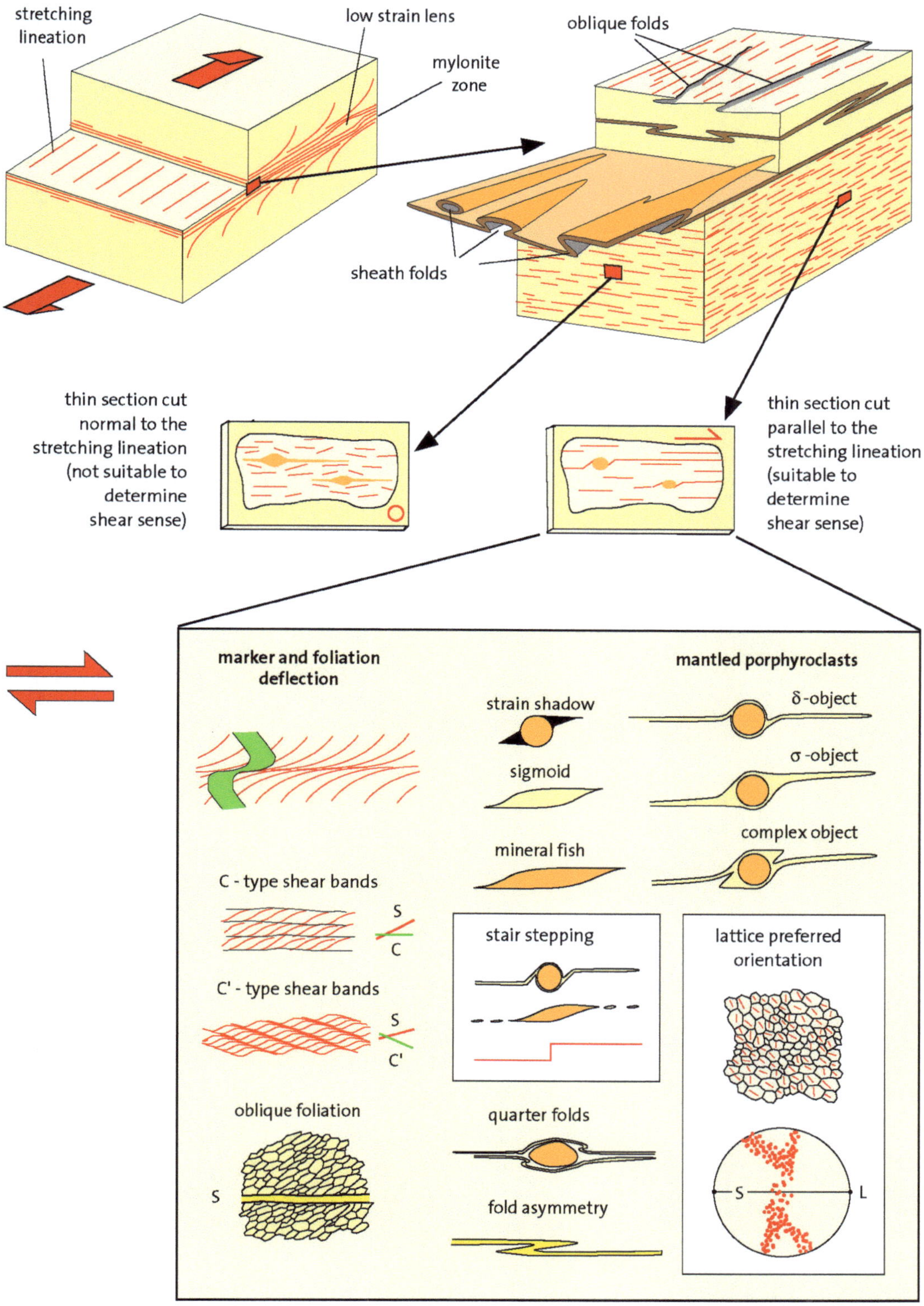

Fig. 9.1 **Top** Schematic diagram showing the geometry of a mylonite zone and the nomenclature used. Notice that shear sense indicators can only be judged correctly in sections parallel to the aggregate or stretching lineation and perpendicular to the mylonitic foliation.
Bottom The most common types of shear sense indicators. See below and Passchier & Trouw (2005) for more detail.

9.1 Core-and-Mantle Structures

It is a common feature in mylonites that porphyroclasts recrystallise along the rim, producing a relatively fine-grained polycrystalline "mantle" around a monocrystalline core. A detailed description of different types of core-and-mantle structures is given in Passchier & Trouw (2005). We would like to emphasize here that the grain size in the mantle is proportional to differential stress during deformation. Since differential stress can be build up at lower temperature to higher levels than at higher temperature, grain size is indirectly proportional to temperature as well. This means that at higher temperatures the mantles usually exhibit a larger grain size.

It is not always easy to distinguish between the recrystallised mantle and the surrounding matrix. Even if the core-and-mantle structure is not very clear, the "stair stepping" criterium can usually be applied to define sense of shear.

Fig. 9.1.1 High-grade mylonite derived from quartz-feldspar gneiss. The coloured grains are biotite. The large porphyroclast is plagioclase and shows a deformed inner part with undulose extinction and gently folded polysynthetic twin lamellae. Especially the lower part shows extensive recrystallisation to strain-free new grains. The main mechanism of recrystallisation is probably subgrain rotation. Notice the ribbon-like structure of quartz in the matrix, bending around the porphyroclast. Shear sense is not easy to determine but is thought to be dextral because of the bending of twin planes in the upper part of the porphyroclast. Três Pontas, Minas Gerais, SE Brazil. Width of view 25 mm. CPL.

Fig. 9.1.2 Fine-grained mylonite to ultramylonite with porphyroclasts of plagioclase and K-feldspar, derived from a leucogranite. The lower porphyroclast composed of plagioclase contains three remnants of a deformed old core with undulose extinction (Fig. 9.1.3), surrounded by a mantle of small recrystallised plagioclase grains. Notice stair stepping to the right of the mylonitic foliation next to the porphyroclasts, indicating dextral shear. Western Australia. Width of view 14 mm PPL.

Fig. 9.1.3 As Fig. 9.1.2. Width of view 14 mm. CPL.

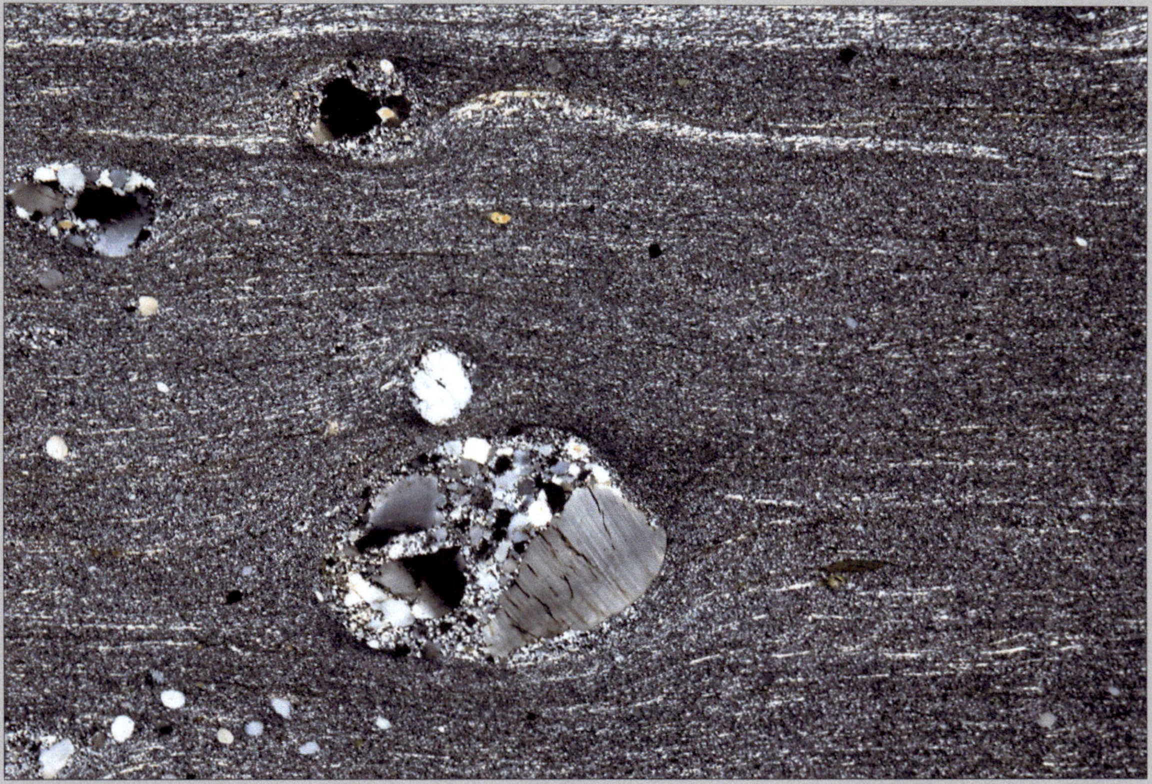

Fig. 9.1.4 Low to medium-grade mylonite with porphyroclasts of K-feldspar, derived from granite. The large porphyroclast just above the center is about 70% recrystallised to small K-feldspar grains, constituting a core-mantle structure. The figure shows a delta structure around this porphyroclast with stair stepping to the right, indicating dextral shear. A fish-like structure at lower left fits this interpretation. Roraima, northern Brazil. Width of view 15 mm. PPL.

Fig. 9.1.5 As Fig. 9.1.4. Width of view 15 mm. CPL.

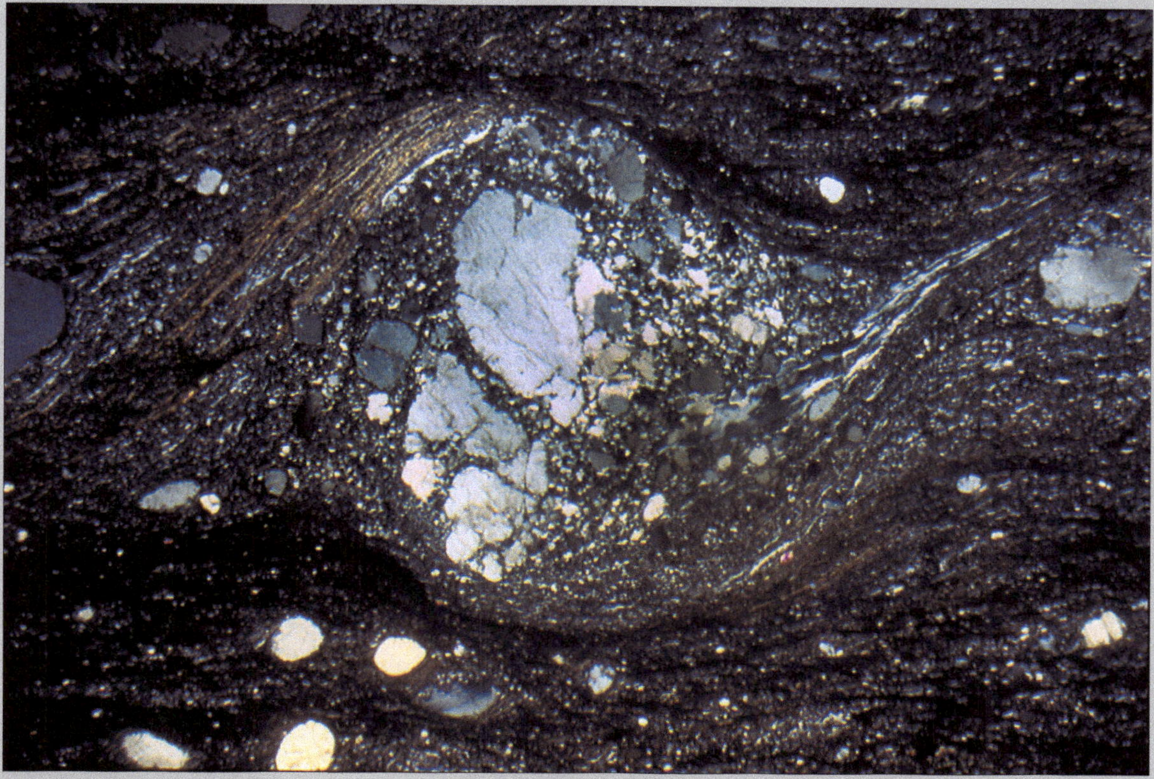

Fig. 9.1.6 Detail of Fig. 9.1.5, showing a clear contrast between the undulose remnants of the deformed K-feldspar grain, surrounded by fine-grained recrystallised K-feldspar. Width of view 10 mm CPL.

Fig. 9.1.7 Low- to medium-grade mylonite with porphyroclasts of K-feldspar (lower left and center right) and plagioclase (lower right and upper right), derived from a granite. The small porphyroclast in the center is partially composed of K-feldspar and partially of plagioclase. Note that some porphyroclasts show advanced recrystallisation (lower right and center) whereas others show very little. A delta-like fabric with stair stepping to the left around the large porphyroclast and C/S structure indicate sinistral sense of shear. Roraima, northern Brazil. Width of view 16 mm. PPL.

Fig. 9.1.8 As Fig. 9.1.7. Width of view 16 mm. CPL.

Fig. 9.1.9 Detail of the lower right porphyroclast of Fig. 9.1.8. About 60 % of the plagioclase is recrystallised and the preserved undulose core is localized in the upper part of the porphyroclast and not in the center. Sense of shear is sinistral as in Fig. 9.1.8. Width of view 4 mm CPL.

Fig. 9.1.10 Low- to medium-grade mylonite derived from granite with a large porphyroclast in the center, composed of zoned plagioclase. Minor recrystallisation along the rim shows initiating core-mantle structure. Notice that the change in darkness in the plagioclase is due to compositional zoning, not to internal deformation. Sense of shear is sinistral as indicated by stair stepping of the mylonitic foliation around the porphyroclast. Leiden collection. Width of view 10 mm. CPL.

Fig. 9.1.11 Detail of Fig. 9.1.10, showing sharp contacts of the recrystallised new grains of plagioclase along the rim of the porphyroclast. Width of view 1.2 mm. CPL.

Fig. 9.1.12 Low- to medium-grade mylonite with porphyroclasts of K-feldspar, composed essentially of biotite, quartz and small feldspar grains. The mylonite is derived from a granite. 10 to 20% recrystallisation along the rims of the porphyroclasts formed a core-and-mantle structure. Sense of shear is dextral, indicated by the upper porphyroclast that developed a sigma structure and the lower one that developed a delta structure, both with stair stepping to the right. Roraima, northern Brazil. Width of view 22 mm. PPL.

Fig. 9.1.13 As Fig. 9.1.12. Width of view 22 mm. CPL.

Fig. 9.1.14 Detail of Fig. 9.1.13, showing "tartan" twinning in the K-feldspar and undulose extinction due to internal deformation, with recrystallisation along the rim. Width of view 6 mm. CPL.

Fig. 9.1.15 Low to medium-grade mylonite, derived from granite, with K-feldspar porphyroclasts. The large porphyroclast below the center shows a clear differentiation between a preserved core and a recrystallised mantle, composed of small grains of K-feldspar. Shear sense is dextral as can be inferred from the sigma structure around the porphyroclast, with stair stepping to the right. Notice also an opaque fish structure in the upper right hand part of the photomicrograph, indicating dextral shear as well. Roraima, northern Brazil.
Width of view 20 mm. PPL.

Fig. 9.1.16 As Fig. 9.1.15. Width of view 20 mm. CPL.

Fig. 9.1.17 Detail of Fig. 9.1.16. The dark core shows vague undulose extinction and makes a sharp transition to the fully recrystallised rim. Width of view 5 mm. CPL.

Fig. 9.1.18 K-feldspar porphyroclast from the same mylonite as shown in Figs. 9.1.15 and 16. Sense of shear is dextral as indicated by the sigma shape of the porphyroclast with stair stepping to the right. Notice a concentration of recrystallised grains in the upper left and lower right quadrants of the porphyroclast, consistent with areas where the porphyroclast suffered its principal shortening. Roraima, northern Brazil. Width of view 13 mm. PPL.

Fig. 9.1.19 As Fig. 9.1.18. Width of view 13 mm. CPL.

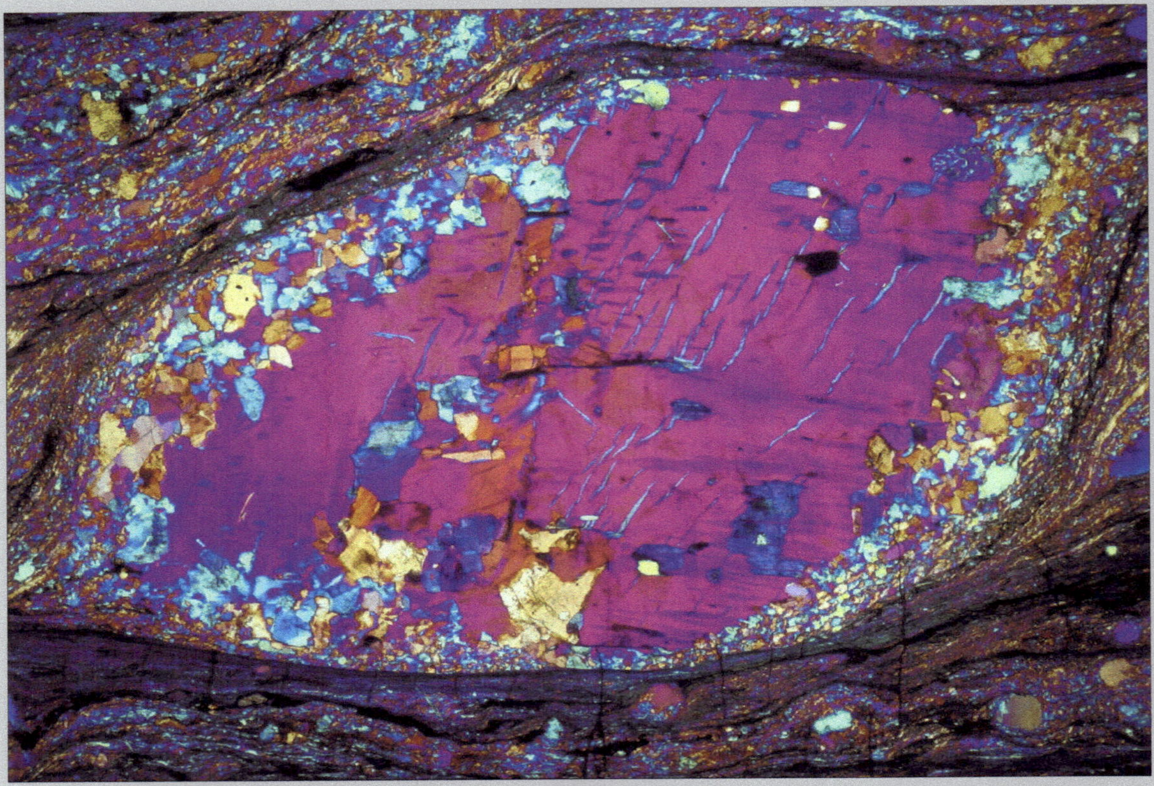

Fig. 9.1.20 Detail of Fig. 9.1.19. This image with gypsum plate shows clearly the contrast in lattice orientation between the core and the recrystallised grains in the rim. Width of view 5 mm. CPL with gypsum plate.

Fig. 9.1.21 Detail of Fig. 9.1.19, showing undulose extinction of the core and recrystallisation in the rims. Width of view 4 mm. CPL.

9.2 Delta and Sigma Structures

Ideal delta and sigma structures (Figs 9.2.1 and 9.2.2) are relatively rare in our experience since they are best developed when they occur isolated in a fine-grained matrix, a not too common occurrence in mylonites. However, similar and transitional structures are common, especially in low-grade mylonites. Transitions to structures with asymmetric strain shadows are also common and in many cases it is hard to distinguish between wings composed of recrystallised material derived from the porphyroclast, and the matrix. In those cases, stair stepping is usually the most reliable shear sense indicator.

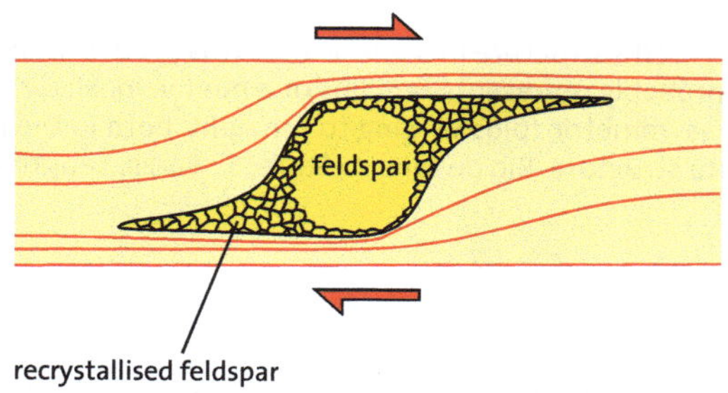

Fig. 9.2.1 Idealised sigma structure. Notice that the wings are composed of recrystallised feldspar derived from the porphyroclast.

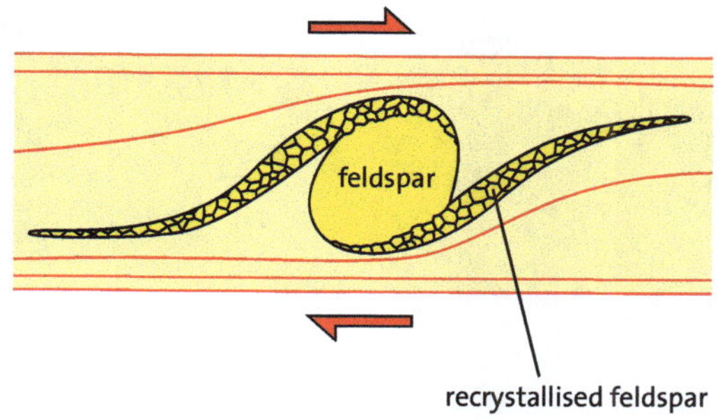

Fig. 9.2.2 Idealised delta structure.

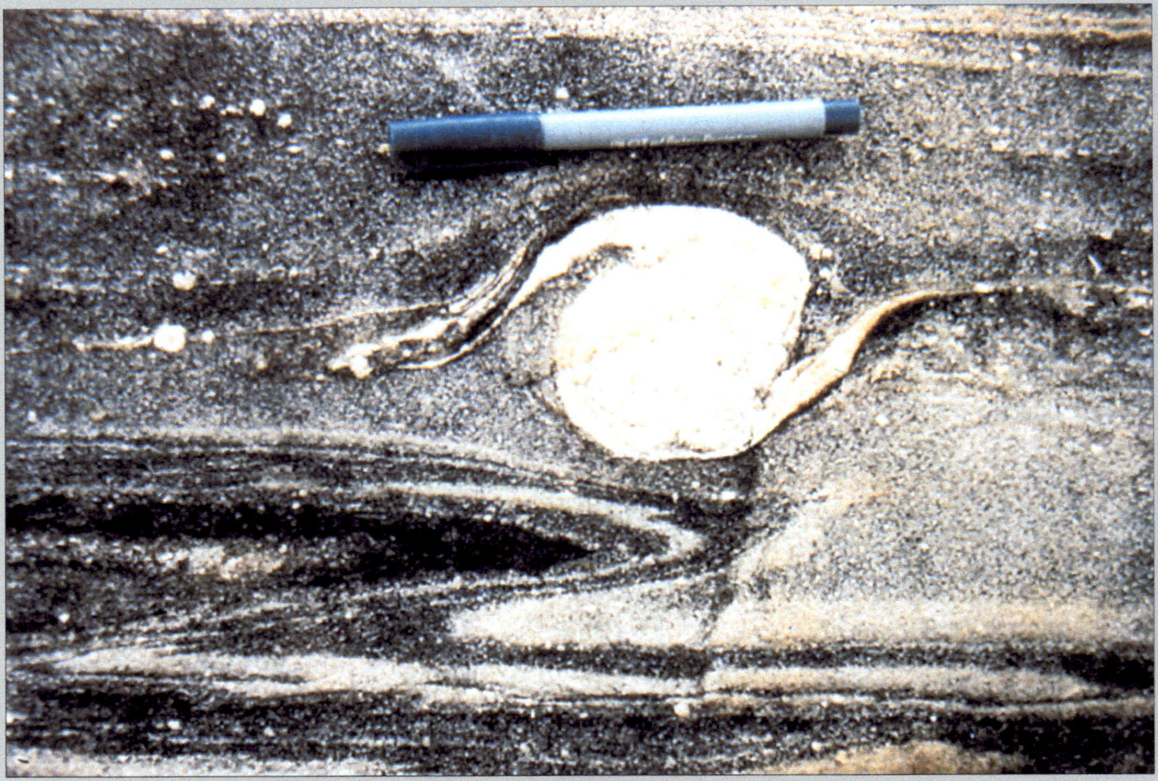

Fig. 9.2.3 Delta structure in a gneiss that was probably once a mylonite. The porphyroclast is composed of feldspar and probably represents remains of a melt vein. Shear sense is dextral as indicated by stair stepping to the right and asymmetric folds verging to the right, both in the matrix below the porphyroclast, and in the left wing of the delta structure. Rio de Janeiro State, SE Brazil. Length of pen 13 cm.

Fig. 9.2.4 Delta structure in a low-grade ultramylonite, derived from a granite. The porphyroclast is K-feldspar and the matrix is dark because it is rich in very fine-grained biotite. Shear sense is dextral as indicated by the stair stepping to the right across the delta structure. Saint-Barthélemy Massif, French Pyrenees. Width of view 2 mm. PPL.

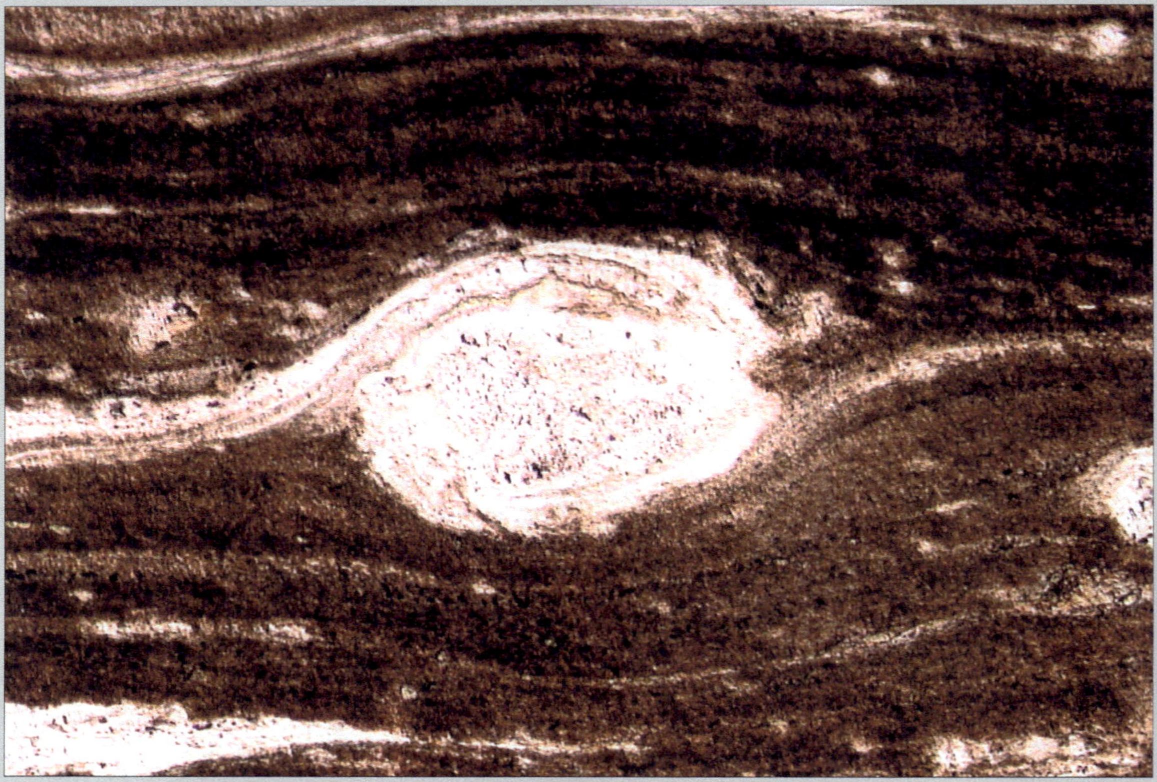

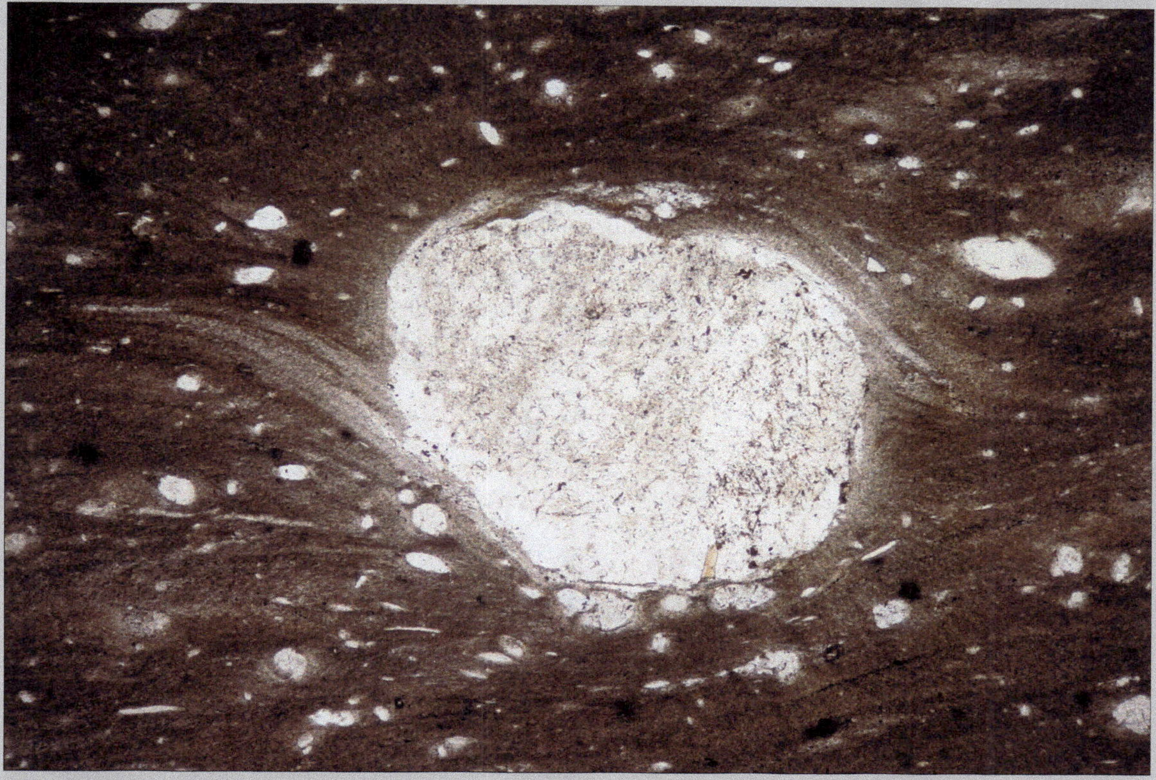

Fig. 9.2.5 Delta structure around K-feldspar porphyroclast in a low to medium-grade mylonite derived from a paragneiss. The dark matrix is rich in very fine-grained biotite. The wings of this delta structure are not composed of recrystallised K-feldspar but rather represent flow structure in the matrix. Shear sense is sinistral as shown by the stair stepping to the left. South Island, New Zealand. Width of view 3.5 mm PPL.

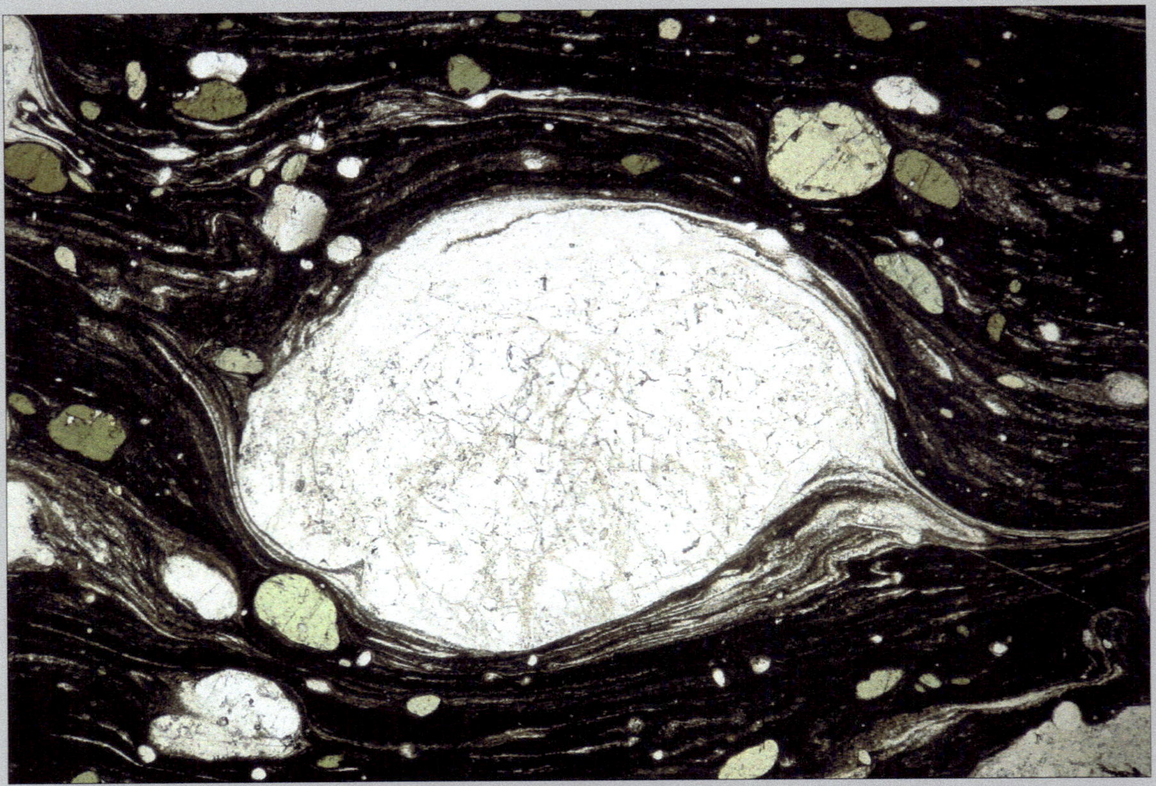

Fig. 9.2.6 Mylonite derived from an amphibolite with porphyroclasts of plagioclase and hornblende. The large porphyroclast in the center is plagioclase with a strongly undulose core, surrounded by a very fine-grained recrystallised mantle, containing some quartz. The delta structure indicates sinistral sense of shear, confirmed by stair stepping and quarter folds in the lower right quadrant. Notice that hornblende forms naked clasts that do not define sense of shear. Mt. West, Western Australia. Width of view 4 mm PPL.

Fig. 9.2.7 As Fig. 9.2.6. Width of view 4 mm CPL.

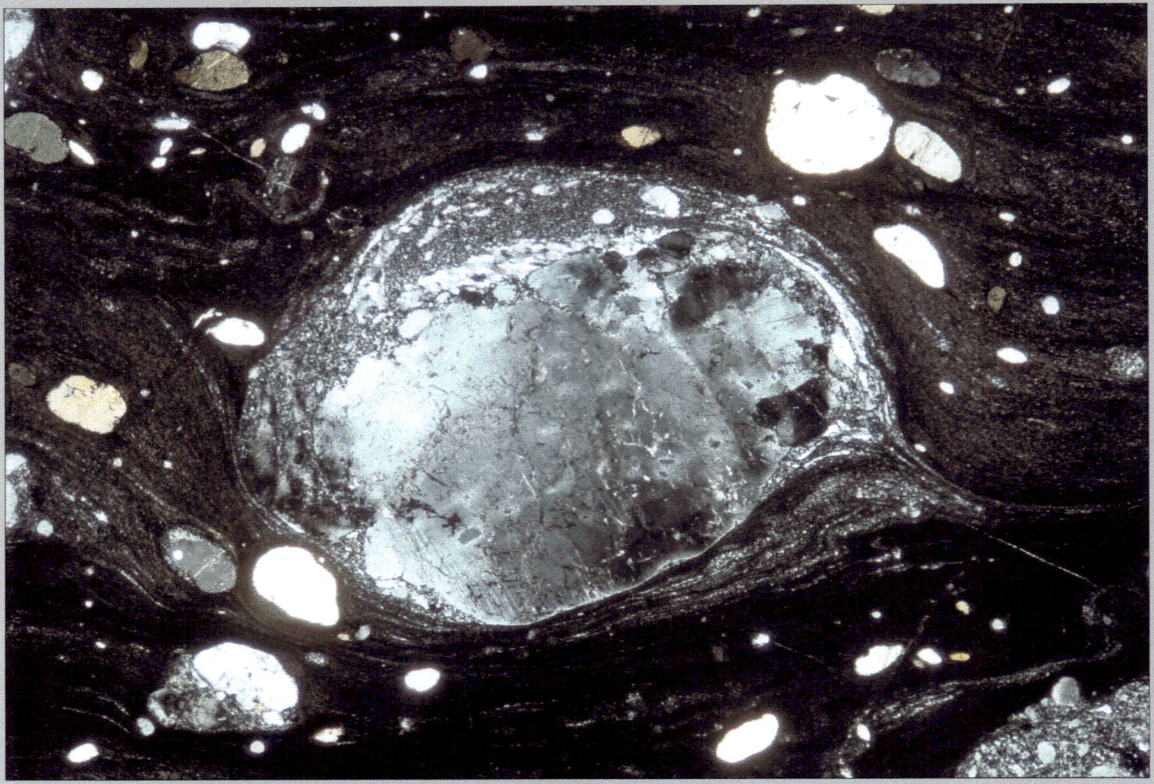

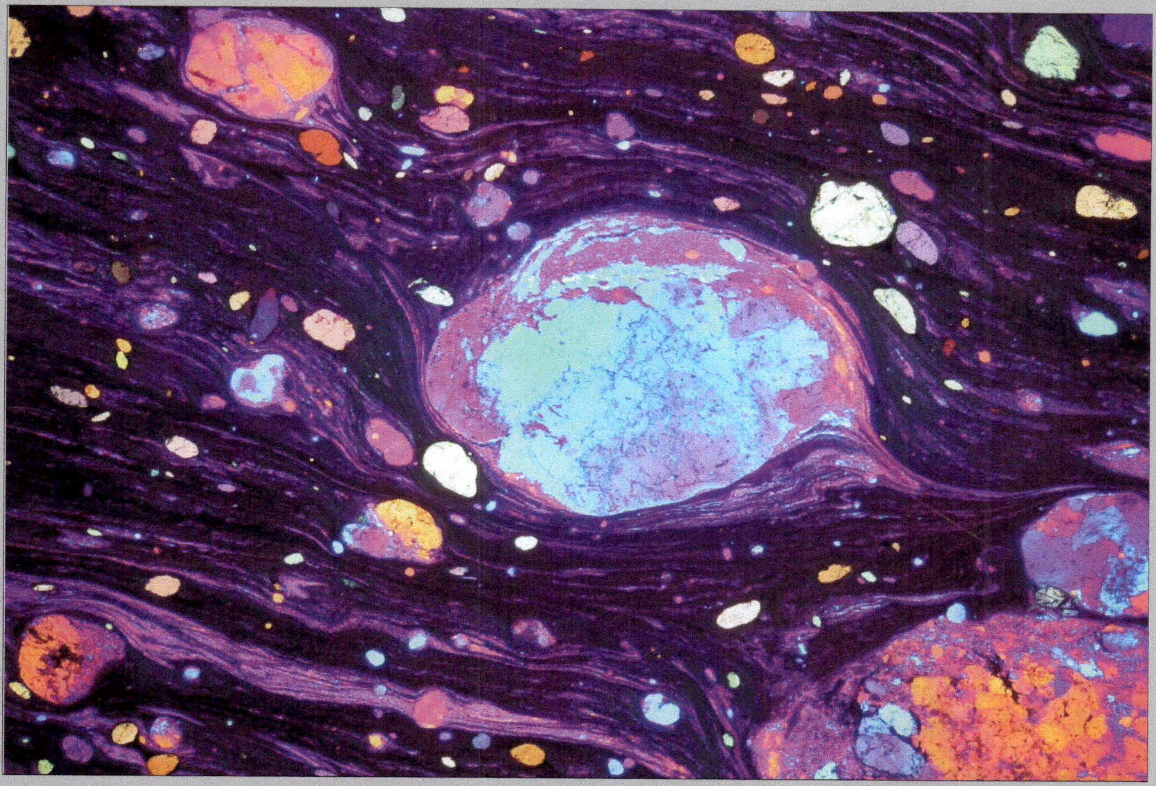

Fig. 9.2.8 As Fig. 9.2.6. Width of view 5 mm. CPL with gypsum plate.

Fig. 9.2.9 Delta clast from the same thin section as Fig. 9.2.6, showing recrystallisation to very fine-grained plagioclase, especially in the upper part. The wing on the left and lower part is partly composed of strongly deformed quartz. Shear sense is sinistral. Width of view 1.2 mm. PPL.

Fig. 9.2.10 As Fig. 9.2.9. Width of view 1.2 mm. CPL.

Fig. 9.2.11 Delta clast composed of plagioclase indicating sinistral sense of shear, from the same thin section as Fig. 9.2.6. At first sight the naked clasts of hornblende do not seem to indicate shear sense but careful examination of the asymmetry in their strain shadows shows that they are coherent with sinistral sense of shear. Width of view 1.2 mm. PPL.

Fig. 9.2.12 Delta structure from the same thin section as Fig. 9.2.6. The porphyroclast is plagioclase and the sense of shear is sinistral. Width of view 1.2 mm. PPL.

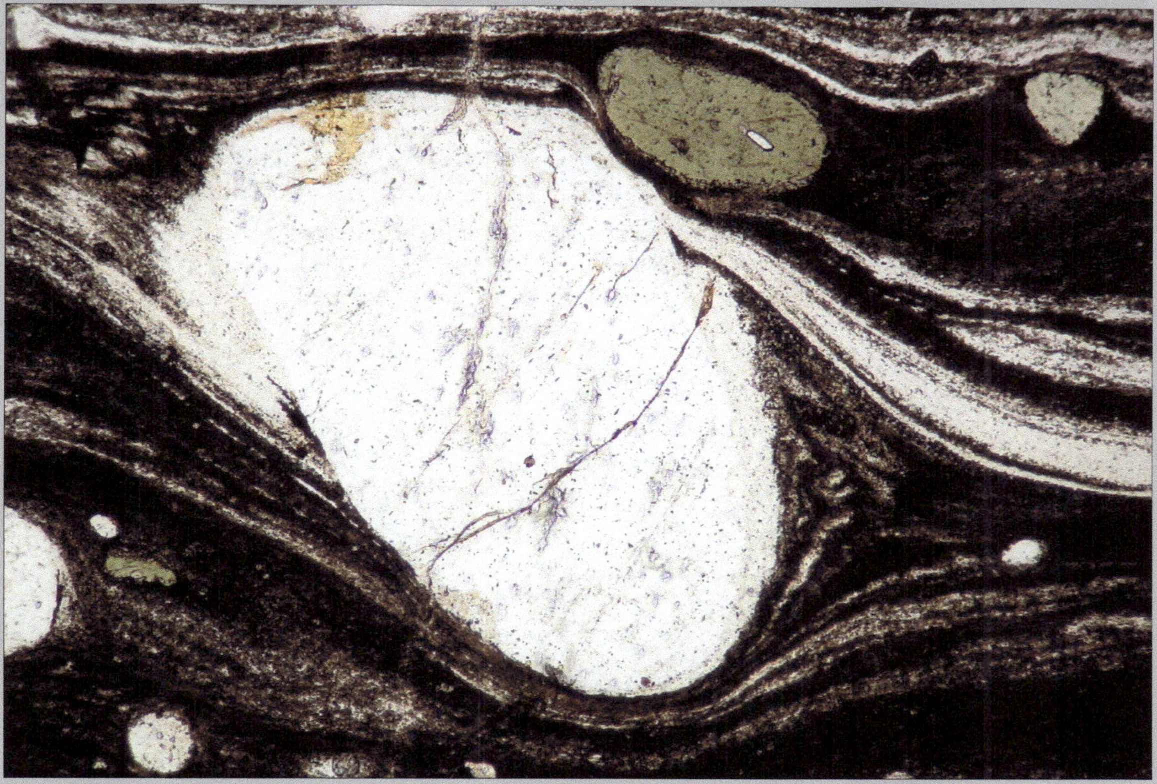

Fig. 9.2.13 Delta structure from the same thin section as Fig. 9.2.6. The accentuated stair stepping shows the sinistral sense of shear. The porphyroclast is plagioclase. Width of view 1.2 mm. PPL.

Fig. 9.2.14 Low- to medium-grade mylonite, derived from paragneiss. The porphyroclasts are composed of K-feldspar, plagioclase and muscovite. The large K-feldspar porphyroclast in the center is a sigma structure with stair stepping to the left indicating sinistral sense of shear. Mica fish at lower left and asymmetric folds left of the central porphyroclast also indicate sinistral shear sense. Quartz in the white bands just above the central porphyroclast is recrystallised to small equidimensional grains (Fig. 9.2.15). Pernambuco, NE Brazil.
Width of view 5 mm. PPL.

Fig. 9.2.15 As Fig. 9.2.14. Width of view 5 mm. CPL.

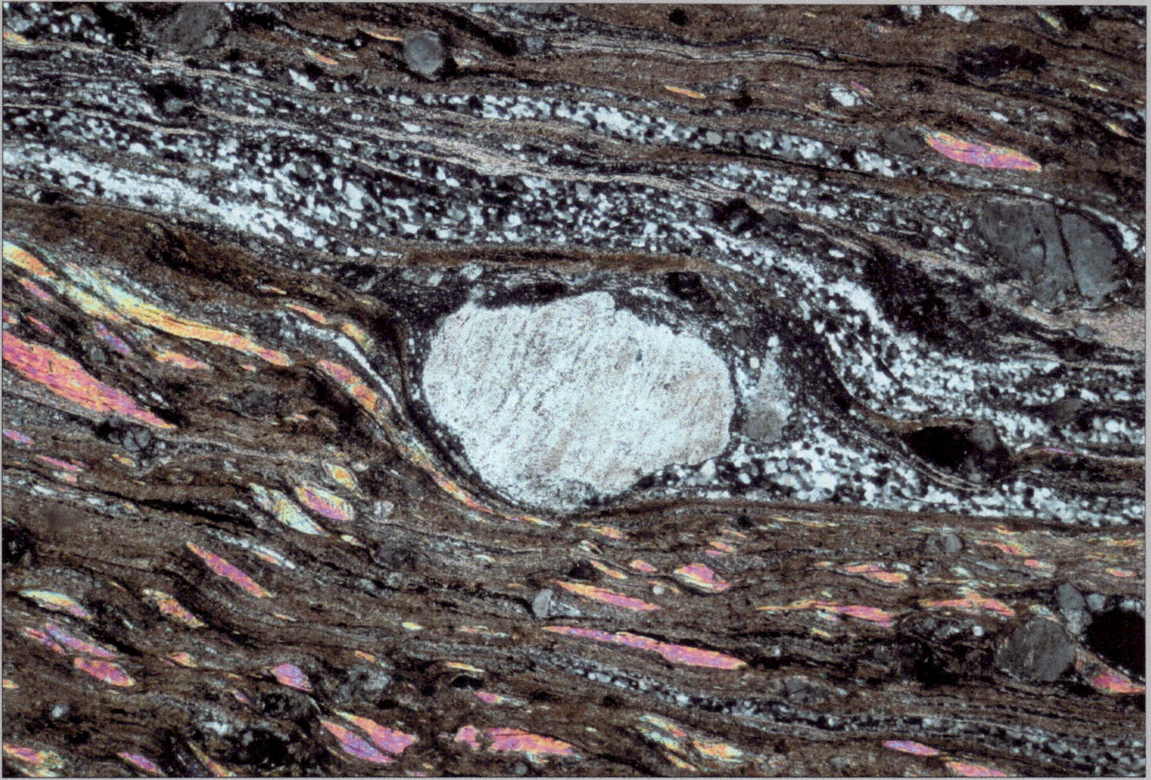

Fig. 9.2.16 Sigma-like structure around the central porphyroclast, which is plagioclase with some biotite attached to the left. There is no significant recrystallisation in the feldspar but the flux structure of the mylonitic foliation forms asymmetric strain shadows that define the sigma structure. Shear sense is sinistral defined by the stair stepping to the left. South Island, New Zealand. Width of view 10 mm. PPL.

Fig. 9.2.17 Sigma- and/or fish-like structure around K-feldspar porphyroclast in an ultramylonite. Note the partial recrystallisation of the K-feldspar into very fine-grained new grains. Sense of shear is dextral as is clear from the stair stepping across the structure. Saint Barthélemy Massif, French Pyrenees. Width of view 2 mm. CPL.

Fig. 9.2.18 Low-grade mylonite derived from granite with porphyroclasts of plagioclase, K-feldspar and mica. The sigma shape of the central porphyroclast is mainly due to internal deformation, not to recrystallisation. Stair stepping to the left shows sinistral sense of shear, confirmed by small asymmetric folds below the center. South Island, New Zealand. Width of view 10 mm. PPL.

Fig. 9.2.19 Low to medium-grade mylonite derived from paragneiss. The central porphyroclast of K-feldspar shows a structure transitional between delta and sigma. Shear sense is sinistral indicated by stair stepping and some mica fish in the upper part of the photomicrograph. Pernambuco, NE Brazil. Width of view 4 mm. PPL.

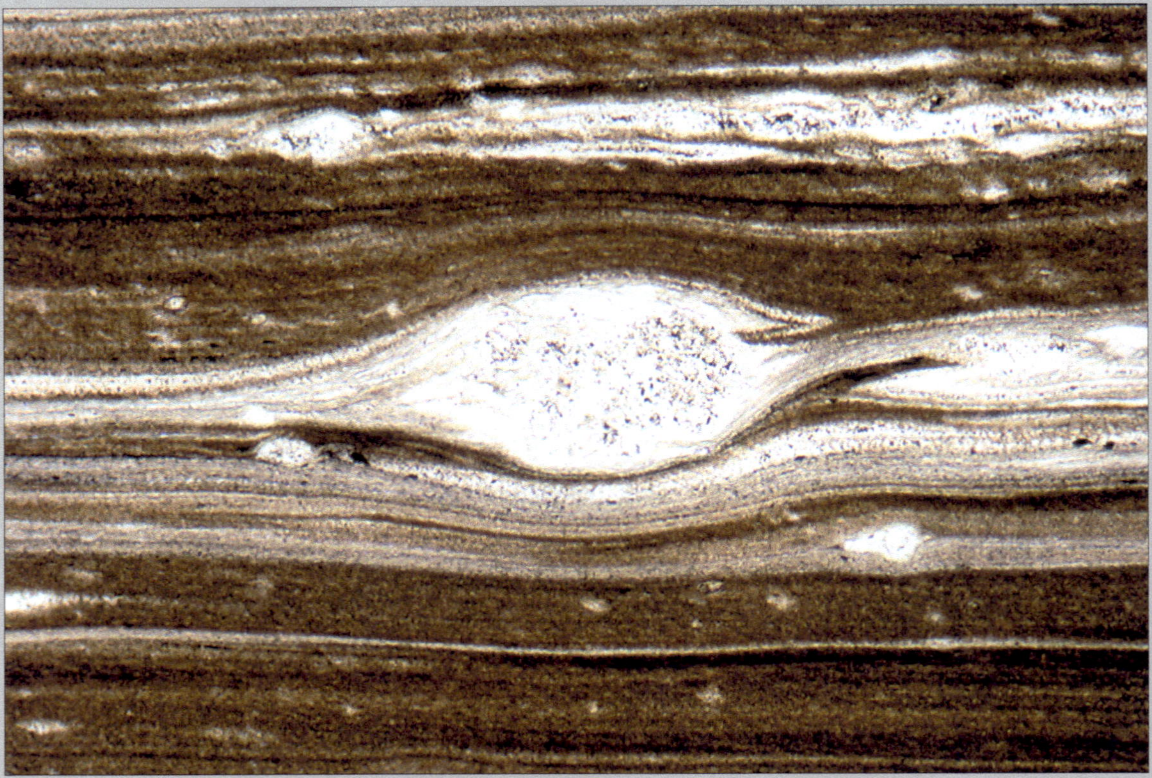

Fig. 9.2.20 Low-grade ultramylonite derived from a sillimanite-bearing paragneiss. The porphyroclast in the center is composed of K-feldspar. It shows a complex core-mantle structure, a transition between sigma and delta structures. Shear sense is dextral. Saint Barthélemy Massif, French Pyrenees. Width of view 3 mm. PPL.

9.3 C/S fabric and C´-type shear bands

C/S and C′planes of shear band cleavage, shown in an idealised form in Figs. 9.3.1 and 9.3.2, are sometimes confused and erroneous interpretation may lead to the wrong sense of shear. S planes are inclined "backwards" with respect to the plane of the shear zone and C′ planes are inclined "forewards", both with similar angles of about 10° to 45°. S planes tend to be penetrative, curved and preserved in less deformed lenses that survive between micro-shear zones or shear bands. The latter ones define C or C′ planes that are usually dark, straight and spaced. The difference between C and C′planes is only their orientation with respect to the main mylonitic foliation or the shear zone itself. C planes are parallel and C′planes are inclined, usually about 30°, but sometimes more. In several cases it could be demonstrated that C′-type shear bands developed later than C-type ones, usually still related to the same mylonitising event.

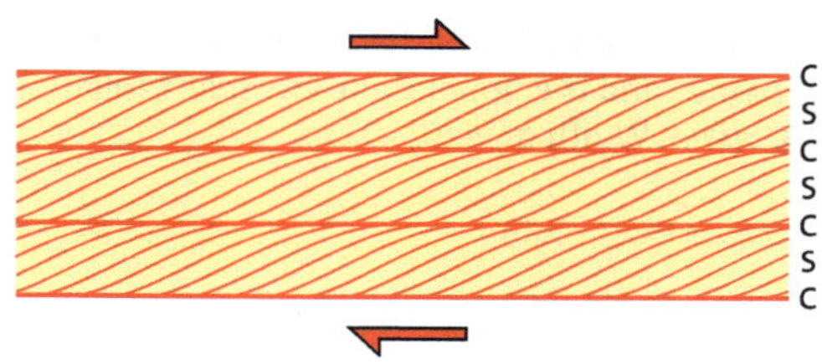

Fig. 9.3.1 Idealised C/S fabric or C/S structure

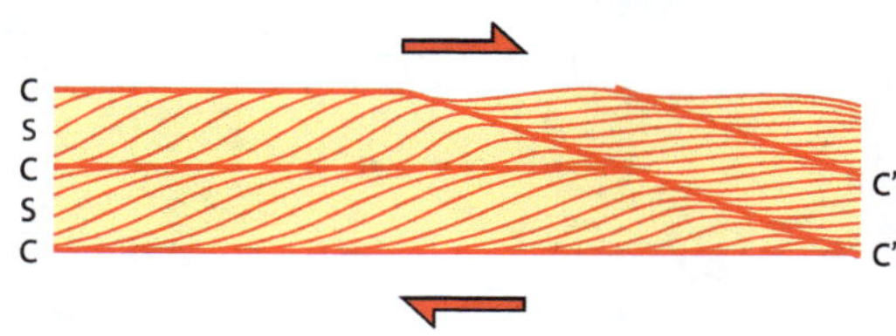

Fig. 9.3.2 Idealised C/S fabric with C´-type shear bands developed at the right hand side.

Fig. 9.3.3 Photograph of a polished rock slab, showing a mylonite derived from a granite. The horizontal black lines are C planes and the upper right-lower left foliation represents S planes. Sense of shear is dextral. Salamanca, Spain. Width of view 10 cm.

Fig. 9.3.4 Photograph of the same polished rock slab with a C' shear band (upper left to lower right). Shear sense is dextral. The C/S fabric is visible in the upper right and lower left corners. Width of view 10 cm.

Fig. 9.3.5 Protomylonitic micaceous quartzite with horizontal shear bands (C planes) in which an oblique foliation (S planes), trending upper left to lower right, is just visible. The sense of shear is sinistral. Western Australia. Width of view 12 mm CPL.

Fig. 9.3.6 Mylonitic paragneiss with biotite and chlorite, showing a C' shear band (upper left to lower right), deflecting the horizontal mylonitic foliation, indicating dextral sense of shear. Southern Minas Gerais State, SE Brazil. Width of view 10 mm. PPL.

Fig. 9.3.7 Mylonite derived from a garnet kyanite paragneiss. An inclined shear band (C') is developed from upper left to lower right in the lower biotite-rich part. The sense of shear is dextral. Marsfjällen, Västerbotten, Sweden. Width of view 7 mm. PPL.

Fig. 9.3.8 This mylonite, derived from a granite, provides a nice example of a C/S fabric. The S planes are inclined to the right and the C planes are slightly inclined to the left. The sense of shear is sinistral. Rio Grande do Sul, southern Brazil. Width of view 13 mm. PPL.

Fig. 9.3.9 As Fig. 9.3.8. Width of view 13 mm. CPL.

Fig. 9.3.10 Mylonitised granite showing a C/S fabric with C planes subhorizontal to slightly inclined to the right and curved S planes trending top right to lower left. The sense of shear is dextral. Western Australia. Width of view 15 mm. CPL.

Fig. 9.3.11 Low-grade mylonite derived from a granitodiorite. The upper part is a protomylonite with a developing mylonitic foliation (S planes), inclined to the left. It is cut by a mica rich band at left, inclined to the right, which can be interpreted as a C´ shear band. The protomylonitic upper part grades into a dark band of ultramylonite with a strong horizontal mylonitic foliation (S and C planes become parallel by progressive rotation of S towards C). Below this is a band of mylonite with some S planes preserved, and then another ultramylonite. The inclined foliation in the lowest part of the section is again an S foliation being almost parallel to the C planes in the ultramylonite above. The sense of shear is dextral. Saint-Barthélemy Massif, French Pyrenees. Width of view 20 mm. PPL.

Fig. 9.3.12 As Fig. 9.3.11. Width of view 20 mm. CPL.

Fig. 9.3.13 Mylonite derived from a schist. The S planes are inclined to the right and few dark C planes are sub-horizontal. The sense of shear is sinistral. Heliodoro, southern Minas Gerais State, SE Brazil. Width of view 12 mm. PPL.

Fig. 9.3.14 Protomylonite derived from a granite, formed under relatively low-grade conditions. The horizontal C planes are marked by a concentration of mica. S-planes are inclined to the left. The sense of shear is dextral. Rio Grande do Sul, southern Brazil. Width of view 23 mm. CPL.

Fig. 9.3.15 Mylonitised granite with typical C/S fabric. The C planes are discontinuous and subhorizontal, and the curved S planes are inclined to the right. In the top layer, a C´ shear band, inclined to the left, is present. The sense of shear is sinistral. Notice the concentration of mica along the C planes. Salamanca, Spain. Width of view 16 mm. PPL.

Fig. 9.3.16 As Fig. 9.3.15. Width of view 16 mm. CPL.

Fig. 9.3.17 Mylonitic schist with two large garnet porphyroblasts (black) and foliation fish composed of muscovite (Chapter 9.5). The metamorphic circumstances during mylonitisation were medium grade as indicated by the recrystallised quartz (upper right and lower left). The fabric can also be described as a C/S fabric, with dark and discontinuous C planes, inclined to the right, and curved S planes (especially within the foliation fish) inclined to the left. Sense of shear is dextral. Santana do Garambeu, southern Minas Gerais State, SE Brazil.
Width of view 34 mm. CPL.

Fig. 9.3.18 Mylonitised amphibolite with three C´-type shear bands trending upper right to lower left. The sense of shear is sinistral. The white band at lower left is a fracture in the thin section. Cadaquez, NE Spain. Width of view 40 mm. PPL.

Fig. 9.3.19 As Fig. 9.3.18. Width of view 40 mm. CPL.

9.4 Oblique Foliation

Oblique foliation (Fig. 9.4.1) is most commonly developed in quartz-rich parts of gneisses or in micaceous quartzites. It is a steady-state fabric that reflects competition of two processes with opposite effects; a process of deformation tending to produce grains elongated parallel to the main extension directions of the incremental strain ellipsoid, and recrystallisation tending to form new equidimensional grains free of strain. Oblique foliation therefore has the same kinematic significance as S planes in C/S shear bands, lagging behind in orientation with respect to the total strain ellipsoid. Theoretically, oblique foliation should make angles of less than 45° with the main mylonitic foliation, but larger angles have been reported and may be related to transtensional flow in shear zones. Care has to be taken not to confuse oblique foliation with superposition of deformation phases (cf. Figs. 10.12 and 10.13), where the angle can be up to 90°.

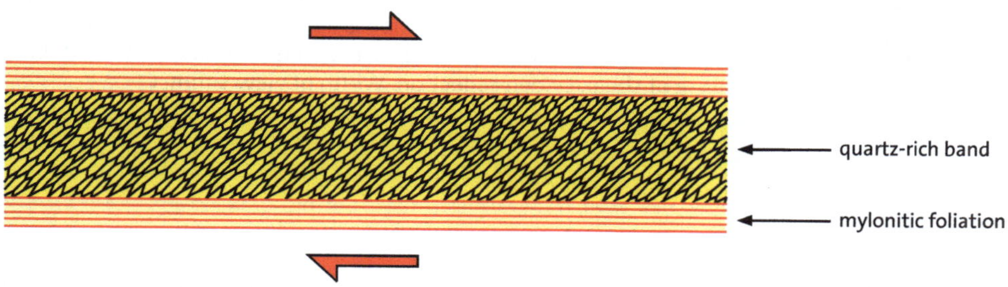

Fig. 9.4.1 Idealised oblique foliation defined by the shape of inequant quartz grains in a quartz-rich band embedded in a mylonite, with the mylonitic foliation as reference frame.

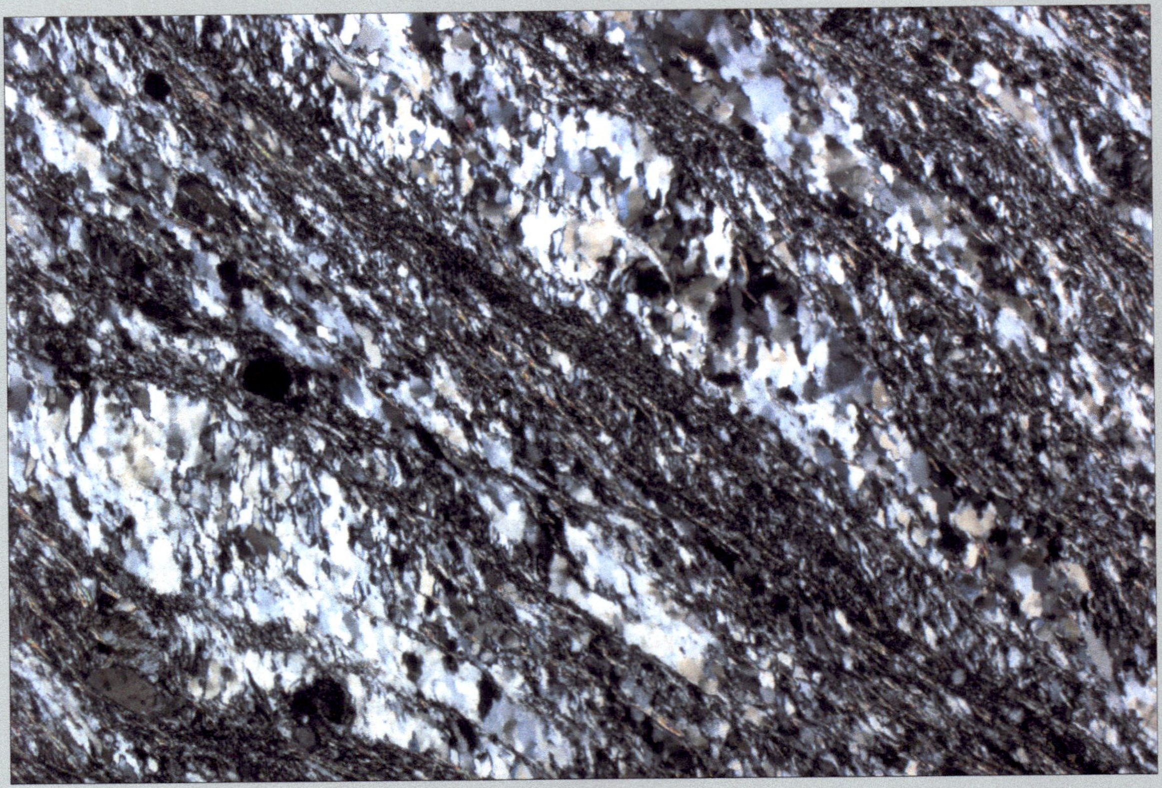

Fig. 9.4.2 Low-grade mylonite derived from a micaceous quartzite. In this case no clear distinction can be made between porphyroclasts and matrix, because no resistant minerals are present to form porphyroclasts. Quartz is strongly deformed by crystal-plastic deformation and no recrystallisation is apparent. The main mylonitic foliation is upper left-lower right, but the shape orientation of elongated quartz grains is almost vertical, defining an oblique foliation indicative of sinistral sense of shear. Leiden collection. Width of view 6 mm. CPL.

Fig. 9.4.3 Detail of Fig. 9.4.2 showing an angle of about 45° between the oblique foliation (almost vertical) and the mylonitic foliation (upper left-lower right). Width of view 3 mm. CPL.

Fig. 9.4.4 Low-grade mylonite derived from granite, with feldspar porphyroclasts in a quartz and biotite-bearing matrix. A quartz-rich band in the lower part shows a well developed oblique foliation, indicating dextral sense of shear. The delta type porphyroclast above the center indicates the same shear sense. Roraima, northern Brazil. Width of view 5 mm. CPL.

Fig. 9.4.5 Fine-grained mylonitic schist rich in quartz, with subhorizontal mylonitic foliation and an oblique foliation defined by elongated quartz (upper right to lower left). The structure indicates dextral sense of shear. Note the absence of porphyroclasts (yet this is not an ultramylonite). Southern Minas Gerais State, SE Brazil. Width of view 5 mm. CPL.

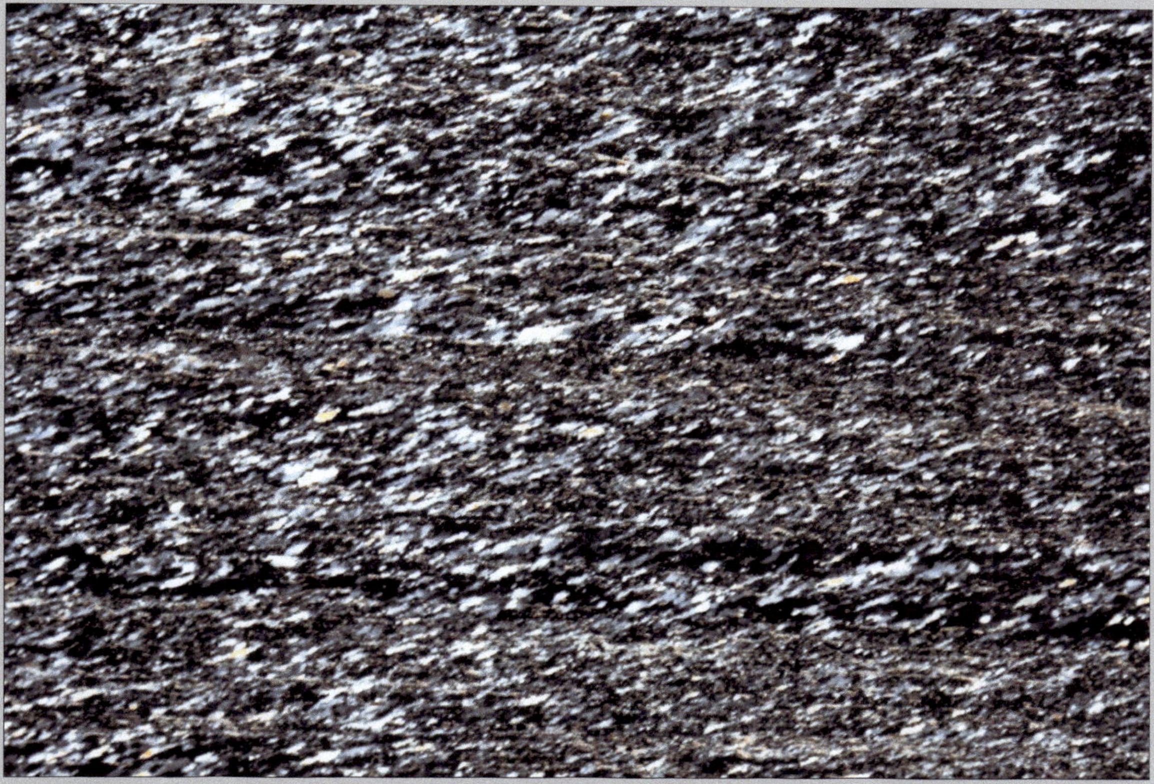

Fig. 9.4.6 Low- to medium-grade mylonitic quartzite with mica fish of muscovite indicating dextral sense of shear by their oblique orientation and stair stepping. Quartz in the matrix is recrystallised and deformed and its preferred shape orientation defines an oblique foliation, inclined to the left, also indicative for dextral sense of shear. Conceição do Rio Verde, southern Minas Gerais State, SE Brazil. Width of view 12 mm. CPL.

Fig. 9.4.7 Low-grade mylonite derived from a granitic rock with porphyroclasts of feldspar in a very fine-grained dark matrix with quartz veins. The quartz in the veins is strongly deformed and partially recrystallised. Its shape defines an oblique foliation, inclined to the right, indicating sinistral shear sense, confirmed by some asymmetric folds. Gosse, Western Australia. Width of view 3 mm. CPL.

Fig. 9.4.8 Low-grade mylonite derived from a granitic rock with quartz-rich band. The quartz is strongly affected by crystal-plastic deformation with only incipient recrystallisation by bulging. The shape of the deformed quartz grains defines an oblique foliation inclined to the right, indicative for sinistral sense of shear. Leiden collection. Width of view 3 mm. CPL.

Fig. 9.4.9 Low-grade ultramylonite derived from granite, with a quartz-rich vein that shows a well developed oblique foliation (upper right lower left) indicating dextral sense of shear. St. Barthélemy Massif, French Pyrenees. Width of view 5 mm. CPL.

9.5 Mineral Fish and Foliation Fish

Mineral fish, especially muscovite fish, have been subdivided into several groups according to their shape (ten Grotenhuis et al. 2003; Fig. 9.5.1) and mechanisms have been proposed to explain these shapes (Passchier & Trouw 2005). However, many shapes are transitional.

Most mineral fish have in common that their longest dimension makes a small angle with the mylonitic foliation, producing stair stepping. This stair stepping is the most reliable shear sense indicator of mineral fish.

The most common mineral to produce fish shapes in mylonites is muscovite, especially in quartz-rich rocks. Biotite fish are relatively rare, probably because they recrystallise more readily than muscovite. Fish shapes of more resistant minerals such as garnet, amphibole, pyroxene, sphene etc. tend to form at high metamorphic grade and usually where these crystals are surrounded by quartz. They may be preserved as relicts of a mylonitic fabric, surrounded by a non-mylonitic rock composed to a large extent of recrystallised quartz, grown to large grainsize.

Foliation fish (Fig. 9.5.2) are polycrystalline lenses, exhibiting similar shapes as mineral fish. The examples shown in this section are all composed of muscovite in a quartz-rich matrix. Foliation fish may grade into C/S fabrics.

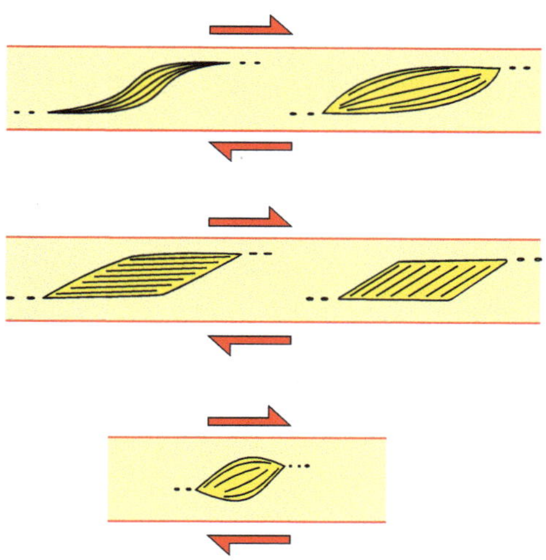

Fig. 9.5.1 Five types of mica fish, commonly encountered in nature.
For further discussion see ten Grotenhuis et al. (2003) and Passchier & Trouw (2005).

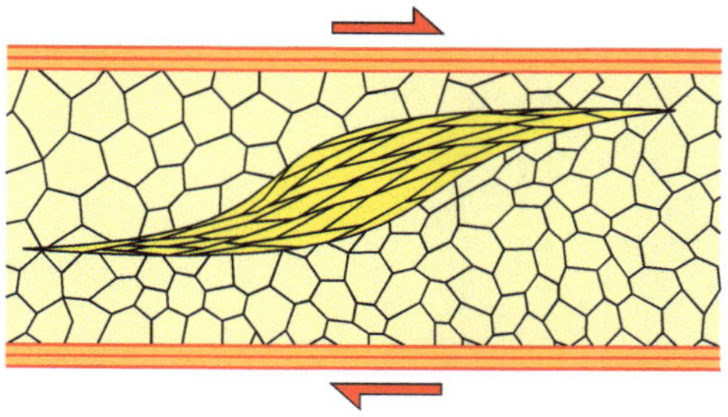

Fig. 9.5.2 Idealised foliation fish composed of mica crystals embedded in a quartzite matrix.

Fig. 9.5.3 Low-grade ultramylonite derived from paragneiss with muscovite fish indicating dextral sense of shear. The brownish matrix is rich in very fine-grained biotite. St. Barthélemy Massif, French Pyrenees. Width of view 6 mm. CPL.

Fig. 9.5.4 Low- to medium-grade mylonite derived from micaceous quartzite with muscovite fish in a recrystallised quartz matrix. Notice different shapes of the fish and undulose extinction. The asymmetric shape and stair stepping of the fish indicate dextral sense of shear. A weak oblique fabric in the quartzitic matrix also indicates dextral shear. Southern Minas Gerais State, SE Brazil. Width of view 12 mm. CPL.

Fig. 9.5.5 Low-grade mylonite derived from schist with abundant muscovite fish. Many fish have a lenticular shape. The inclined position of most fish indicates dextral sense of shear. Notice the very fine-grained recrystallised quartz in the quartz-rich band above the center. Southern Minas Gerais State, SE Brazil.
Width of view 12 mm. CPL.

Fig. 9.5.6 Medium-grade mylonite derived from micaceous quartzite. Numerous lentiform muscovite fish are inclined upper left-lower right, indicating sinistral sense of shear. In Fig. 9.5.7 the undulose extinction of the fish is apparent and also the recrystallised nature of the quartz, reaching a grain size consistent with medium-temperature conditions. Southern Minas Gerais State, SE Brazil. Width of view 12 mm. PPL.

Fig. 9.5.7 As Fig. 9.5.6. Width of view 12 mm. CPL.

Fig. 9.5.8 Medium-grade mylonite with mica fish, derived from micaceous quartzite. Dextral shear sense is indicated by the inclined position of the fish with respect to the horizontal mylonitic foliation defined by very fine-grained mica fragments apparently torn off the mica fish along C planes. Quartz is recrystallised and has grown by grain boundary migration to irregular shapes. Southern Minas Gerais State, SE Brazil. Width of view 4 mm. CPL.

Fig. 9.5.9 Medium-grade mylonite derived from a quartz-mica schist, Sinistral sense of shear is indicated by the oblique mica fish, inclined to the right. The recrystallised quartz indicates medium-temperature conditions during mylonitisation. Note the undulose extinction of the mica fish. Caxambu, southern Minas Gerais State, SE Brazil. Width of view 5 mm. CPL.

Fig. 9.5.10 Low- to medium-grade mylonite derived from micaceous quartzite. The lens-shaped mica fish indicates dextral sense of shear by its inclined position and stair stepping. Quartz in the matrix is recrystallised to relatively small crystalloblastic new grains that define a weak oblique fabric (upper right – lower left), also indicative of dextral shear. Conceição do Rio Verde, Southern Minas Gerais State, SE Brazil. Width of view 3 mm. CPL.

Fig. 9.5.11 Lens-shaped mica fish with curved lower and upper sides, from the same locality as Fig. 9.5.10. Dextral shear sense is indicated by stair stepping and a relatively strong oblique shape fabric in quartz. Notice incipient grain growth in quartz by grain boundary migration, especially in a dark grain below the fish. Width of view 3 mm. CPL.

Fig. 9.5.12 Lozenge-shaped mica fish with undulose extinction from the same mylonite as Fig. 9.5.10. Note that the mineral cleavage of muscovite is parallel to the sides of the fish. Shear sense is sinistral. A weak oblique fabric in quartz is also present. Width of view 3 mm. CPL.

Fig. 9.5.13 Folded mica fish in mylonite from the same locality as Fig. 9.5.10. Sense of shear is dextral. Notice a weak oblique fabric in the recrystallised polygonal quartz grains. Width of view 3 mm. CPL.

Fig. 9.5.14 Elongated lozenge-shaped mica fish from the same mylonite as Fig. 9.5.10. Notice that the mineral cleavage is parallel to the long sides of the fish, and subparallel to the mylonitic foliation. Although the mica is slightly inclined in a "foreward" rotated position, the tips define stair stepping to the right indicative of dextral sense of shear. The oblique fabric in the recrystallised quartz matrix confirms this interpretation. Width of view 3 mm. CPL.

Fig. 9.5.15 Curious-shaped mica grain that resembles more a bird than a fish. It comes from the same locality as Figs. 9.5.10–14. Shear sense is dextral. Notice a weak oblique fabric and incipient grain boundary migration in the quartz matrix. Width of view 3 mm. CPL.

Fig. 9.5.16 Mica fish with isoclinal microfolds at both tips, from the same mylonite as shown in Figs. 9.5.10–15. The inclined position, stair stepping and vergence of the folds all indicate dextral shear.
Width of view 1.5 mm. CPL.

Fig. 9.5.17 Four mica fish from the same mylonite as shown in Figs. 9.5.10–16. The upper fish shows kink bands in the left hand part and an isoclinal fold in the right hand part. Sense of shear is dextral.
Width of view 6 mm. CPL.

Fig. 9.5.18 Large mica fish with the right hand side apparently thrust over the left hand part. Note that the upper part is defined by an inclined shear band (C′). The angular coloured grains in the lower part are tourmaline. Sense of shear is sinistral. Same locality as Figs. 9.5.10–17. Width of view 6 mm. CPL.

Fig. 9.5.19 Transitional rock between a medium-grade mylonite and a quartz-rich mica schist with a foliation fish composed of a large number of muscovite grains. The shear sense is sinistral as is evident from the inclined position of the fish and the stair stepping. The structure resembles a C/S structure; however, in this case no discrete C planes are developed. Notice the well recrystallised quartz, indicating medium-grade metamorphic conditions during deformation. The micas in the fish structure are also recrystallised to predominantly strain-free new grains. Santana do Garambeu, southern Minas Gerais State, SE Brazil. Width of view 12 mm. CPL.

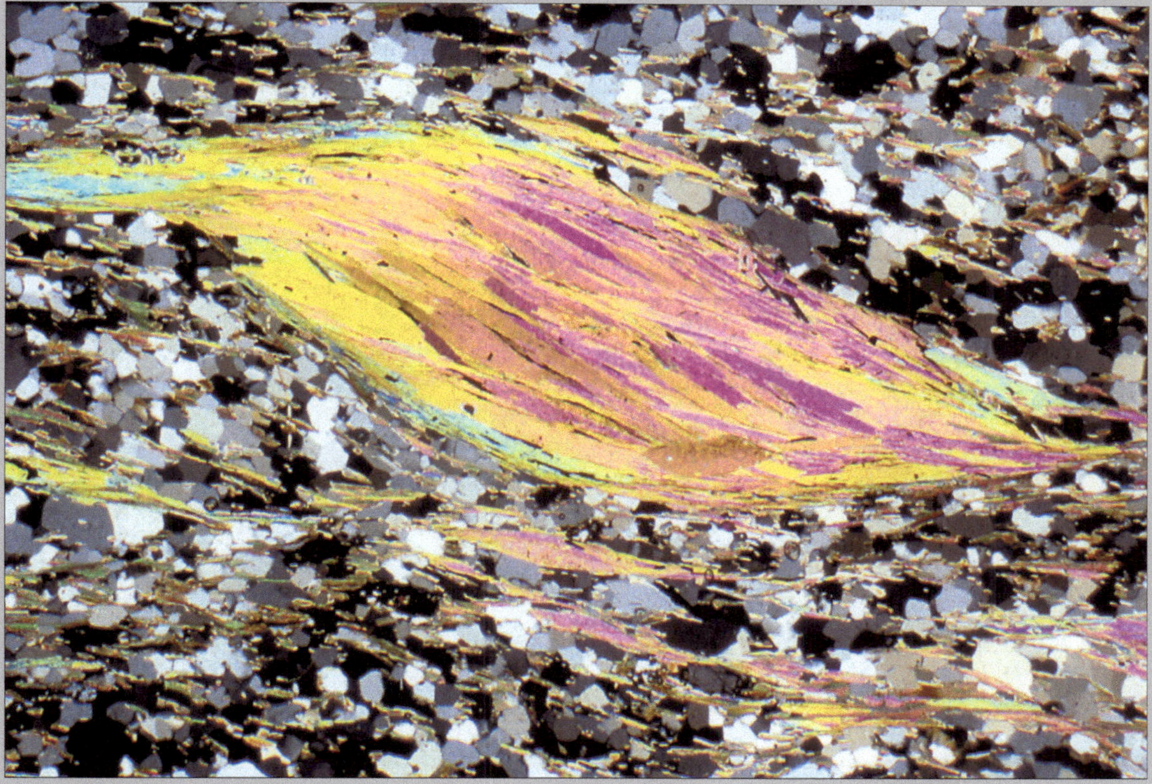

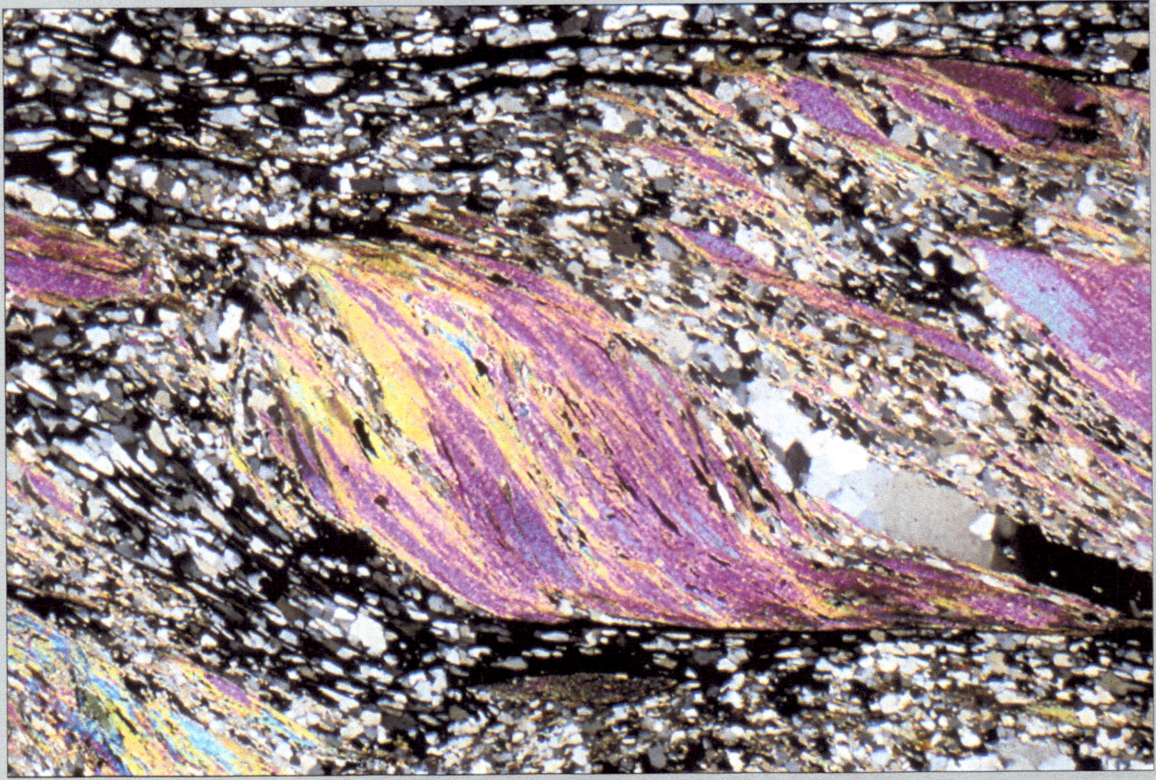

Fig. 9.5.20 Another example of foliation fish from the same region. Shear sense is sinistral. Notice incipient development of C planes (subhorizontal) making this fish transitional to a C/S fabric. Width of view 14 mm. CPL.

Fig. 9.5.21 Garnet-staurolite mica schist with local mylonitic structure. Medium metamorphic grade of deformation is indicated by the well-recrystallised quartz. The foliation fish just above the center contains a staurolite crystal and indicates dextral sense of shear. The quartz vein below the center shows grain growth by grain boundary migration with mica being included in large quartz grains. Santa Rita do Ibitipoca, southern Minas Gerais State, SE Brazil. Width of view 16 mm. CPL.

Fig. 9.5.22 Garnet mica schist with mylonitic structure. Medium metamorphic grade is indicated by recrystallised quartz. The foliation fish indicate dextral sense of shear by their oblique position. Subhorizontal C planes are locally developed making the structure transitional to C/S fabric. The garnets are older than the shear movements, but their shape is possibly modified by them. Santana do Garambeu, southern Minas Gerais State, SE Brazil. Width of view 16 mm. CPL.

Fig. 9.5.23 Quartz-muscovite schist with remnants of foliation fish, being progressively obliterated by late- to post-deformational grain growth both in the micas and in the quartz. Shear sense is sinistral as indicated by the inclined position of the micas just right of the center. Notice grain growth in quartz by grain boundary migration. Luminarias, southern Minas Gerais State, SE Brazil. Width of view 12 mm. CPL.

Fig. 9.5.24 Biotite fish surrounded by quartz, showing internal deformation and recrystallisation along the rims. Shear sense is dextral. Santa Rosa Mylonite zone, California. Width of view 6 mm. PPL.

Fig. 9.5.25 Detail of Fig. 9.5.24, with CPL, showing advanced recrystallisation in the tip and along the rims of the biotite fish. Quartz is also well recrystallised indicating medium metamorphic grade. Width of view 3mm. CPL.

Fig. 9.5.26 Kyanite-bearing micaceous quartzite with lens-shaped kyanite, comparable to mica-fish structure. The oblique position of the kyanite indicates sinistral sense of shear. Quartz in the matrix is recrystallised by grain boundary migration. Conceição do Rio Verde, southern Minas Gerais State, SE Brazil. Width of view 3 mm. CPL.

Fig. 9.5.27 Garnet fish surrounded by large recrystallised quartz grains with several biotite inclusions, indicating dextral shear. This thin section is from a quartz-rich band in a high-grade migmatitic paragneiss, containing sillimanite, K-feldspar and locally cordierite. The rock is not a mylonite, but some mylonitic structures, like this garnet fish, survived in the coarse recrystallised matrix. Recrystallisation and growth of the quartz was by grain boundary migration. Cara de Cão, Rio de Janeiro, Brazil. Width of view 1.5 mm. Polarisers at 45°.

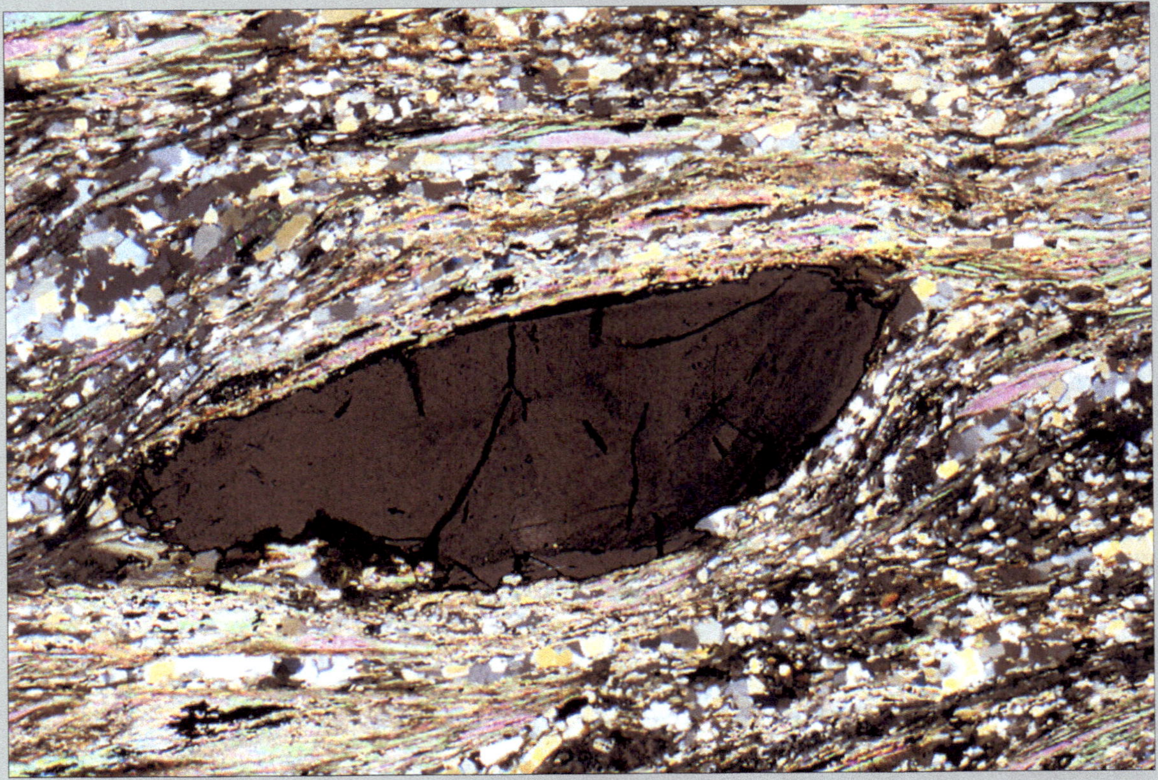

Fig. 9.5.28 Mylonitic garnet mica schist with elongated fish-shaped garnet. The oblique position and the foliation pattern in a C/S-like structure indicate dextral sense of shear. The recrystallised quartz fabric indicates medium metamorphic grade. The shape of the garnet may have been modified by dissolution, preferential reaction and/or mechanical erosion along the upper and lower contacts. Santana do Garambeu, soutern Minas Gerais State, SE Brazil. Width of view 6 mm. Polarisers at 45°.

Fig. 9.5.29 Garnet fish surrounded by large quartz grains that resulted from growth by high-temperature grain boundary migration. The sense of shear is sinistral. Varginha, southern Minas Gerais State, SE Brazil. Width of view 3 mm. Polarisers at 45°.

Fig. 9.5.30 Deformed and recrystallised quartzite containing kyanite (upper left) and ilmenite (opaque). The central fish of ilmenite is completely surrounded by recrystallised quartz and shows sinistral sense of shear by its stair stepping and oblique position. Conceição do Rio Verde, southern Minas Gerais State, SE Brazil. Width of view 1.5 mm. PPL.

Fig. 9.5.31 Low-grade mylonite derived from granite. The central porphyroclast is sphene with an asymmetric shape. This shape was possibly induced by mechanical erosion due to flux of the matrix along the porphyroclast although other processes like dissolution and enhanced reaction along high stress surfaces may have contributed as well. The white porphyroclast to the left is K-feldspar. Shear sense is dextral, indicated by stair stepping across the porphyroclasts. Roraima, northern Brazil. Width of view 2 mm. PPL.

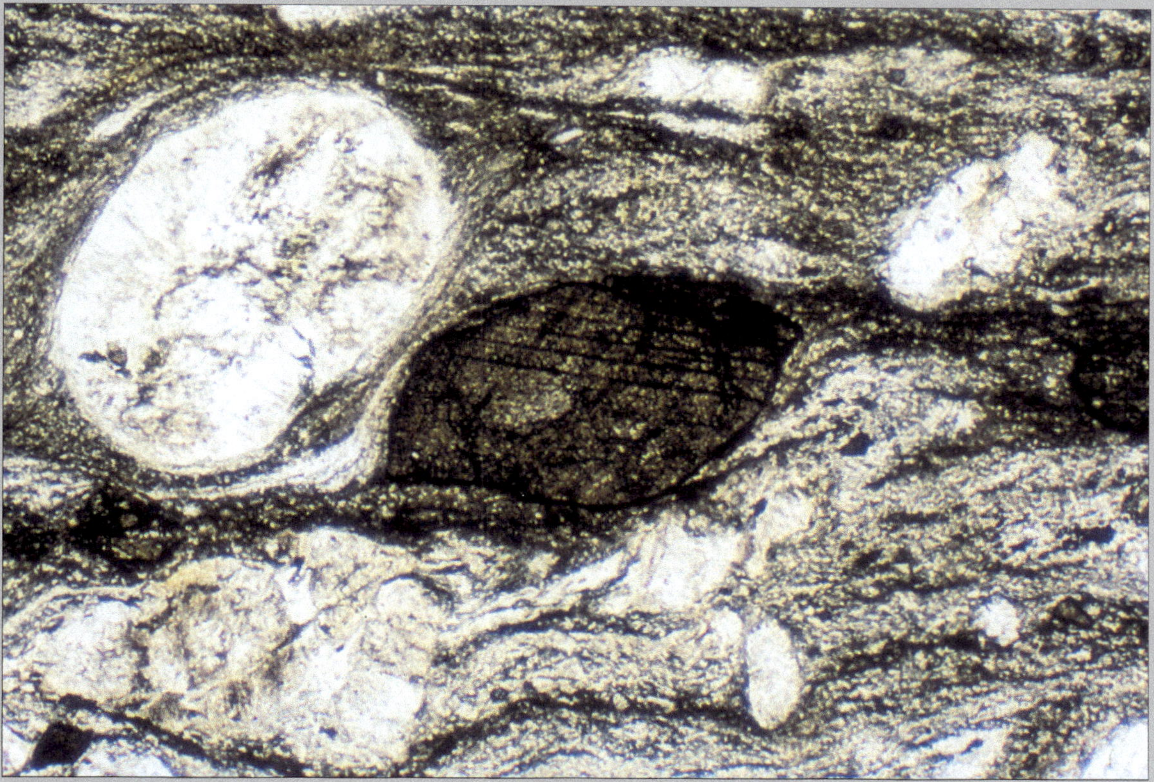

Fig. 9.5.32 Mylonite derived from granulite with granitic composition. The mylonite formed probably under medium-grade conditions as indicated by the small grainsize, strong undulose extinction of feldspar and limited recrystallisation of quartz. The central porphyroclast in Figs. 9.5.32 and 9.5.33 is orthopyroxene that apparently was modified to a fish-like shape by mechanical erosion in flow of the feldspar-rich matrix. Notice the minor shear bands that originate from the upper and lower contacts of the porphyroclasts. Shear sense is dextral, indicated by stair stepping and the oblique position of the porphyroclasts. Caparaó, Espirito Santo State, SE Brazil. Width of view 1.5 mm. CPL.

Fig. 9.5.33 As Fig. 9.5.32. Width of view 2 mm. CPL.

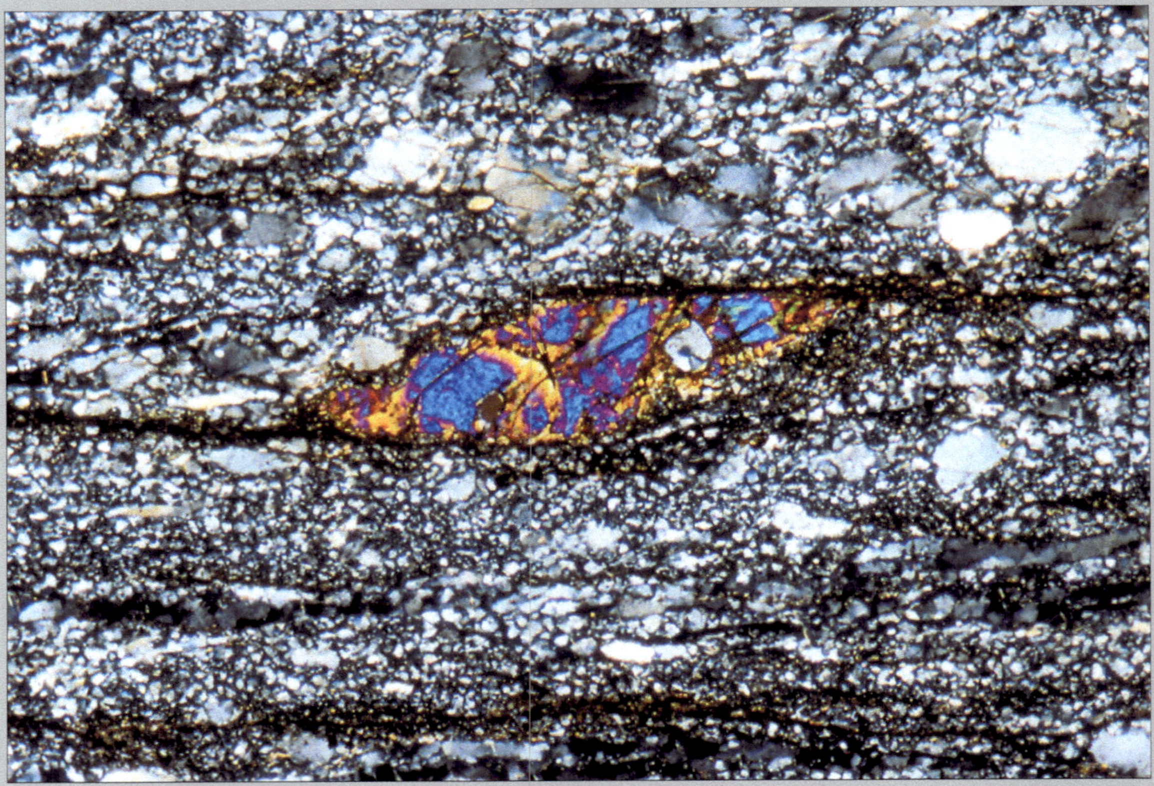

Fig. 9.5.34 Quartzite mylonite with a cluster of cataclased tourmaline grains that form a lens of sigmoidal shape. This shape defines the sense of shear as dextral by its stair stepping and inclined orientation. Conceição do Rio Verde, southern Minas Gerais State, SE Brazil. Width of view 4 mm. PPL.

Fig. 9.5.35 Asymmetric diopside porphyroclast surrounded by large recrystallised quartz. The rock is a high-grade gneiss with kyanite and K-feldspar, indicating relatively high-pressure granulite facies. The fish shaped crystals are interpreted as remnants of a mylonitic structure now mainly masked by recrystallisation and grain growth. It is difficult to determine metamorphic conditions during mylonitisation, but it seems likely that these conditions were medium- to high-grade. The parent rock was a paragneiss with calcsilicate layers. The shape of this diopside indicates dextral sense of shear. Note a biotite inclusion in the upper right part of the diopside that was protected from deformation by the rigid porphyroclast. Varginha, southern Minas Gerais State, SE Brazil. Width of view 1.5 mm. PPL.

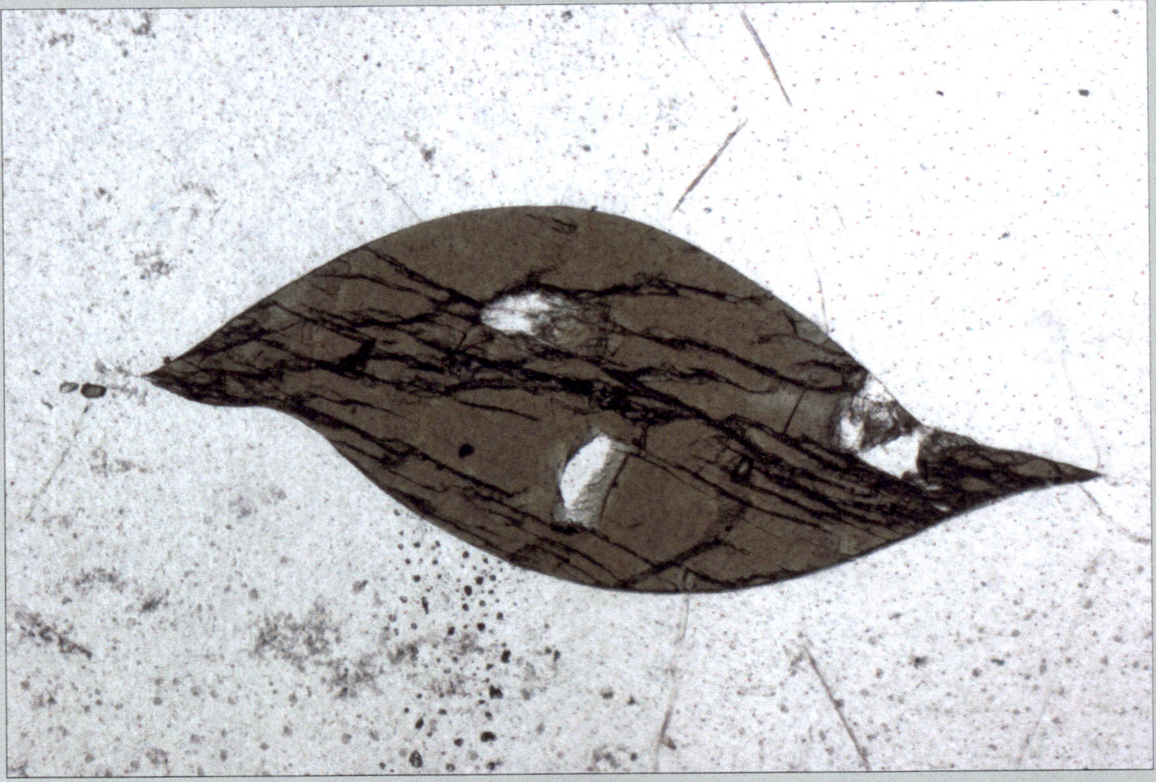

Fig. 9.5.36 The photomicrographs of Figs. 9.5.36 till 9.5.38 show different shapes of hornblende fish, all surrounded by quartz from high-grade paragneisses. The combination of convex and concave shapes is similar to the different shapes of mica fish but there are differences, probably caused by different crystal slip systems. Sense of shear is sinistral in Fig. 9.5.36, and dextral in Figs. 9.5.37 and 38. Varginha, southern Minas Gerais State, SE Brazil. Width of view 1.5 mm. PPL.

Fig. 9.5.37 As Fig. 9.5.36. Width of view 1.5 mm. PPL.

Fig. 9.5.38 As Fig. 9.5.36. Width of view 1.5 mm. PPL.

Fig. 9.5.39 The photomicrographs of Figs. 9.5.39 till 9.5.41 show a variation in shape of plagioclase fish, all surrounded by quartz, recrystallised and grown to large size by grain boundary migration. All indicate sinistral sense of shear. Varginha, southern Minas Gerais State, SE Brazil. Width of view 1.5 mm. CPL.

Fig. 9.5.40 As Fig. 9.5.39. Width of view 1.5 mm. CPL.

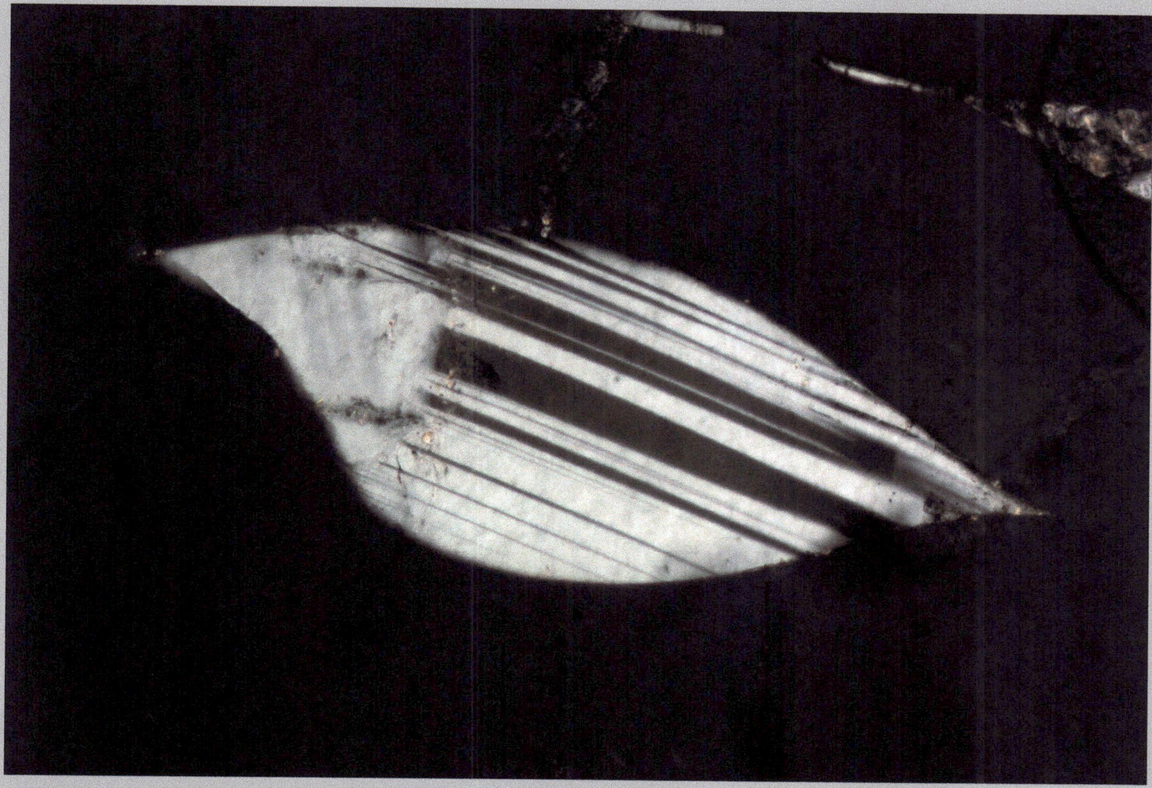

Fig. 9.5.41 As Fig. 9.5.39. Width of view 1.5 mm. CPL.

Fig. 9.5.42 Fish (or humming-bird)-like structure composed of intergrown plagioclase and biotite. The sense of shear is sinistral. The structure is embedded in one large quartz grain; at upper left are three more small plagioclase fish. Varginha, southern Minas Gerais State, SE Brazil. Width of view 1.5 mm. CPL.

Fig. 9.5.43 Mylonitic muscovite schist derived from a rhyolitic tuff that originally contained large phenocrysts of K-feldspar and quartz. K-feldspar grains show a lens or fish shape, with extensive substitution by myrmekite along the upper and lower contacts (Figs. 9.5.44 and 9.5.45). The shape and orientation of the lens or fish indicates sinistral sense of shear, confirmed by the asymmetric distribution of myrmekite. Southeast Minas Gerais State, SE Brazil. Width of view 20 mm. PPL.

Fig. 9.5.44 As Fig. 9.5.43. Width of view 20 mm. CPL.

Fig. 9.5.45 Detail of Fig. 9.5.44, showing partial substitution of K-feldspar by myrmekite along the lower and upper contacts. In the tip of the porphyroclast (upper left) this substitution is not well developed. Width of view 5 mm. CPL.

Fig. 9.5.46 Mylonitic quartz mica schist with “porphyroclasts” of quartz in a fine-grained mica-rich matrix. The large grains of quartz are remnants of quartz phenocrysts in a rhyolitic tuff, the parent rock of this deformed end product. The large quartz grains resisted more than the matrix and remained as porphyroclastic lenses in the sheared matrix. The strong undulose extinction in quartz indicates crystal-plastic deformation with limited bulging recrystallisation (Fig. 9.5.47), typical of low-grade metamorphic conditions. Sense of shear is sinistral indicated by the inclined shape of the quartz lenses and incipient C´ shear bands slightly inclined to the left. Southeast Minas Gerais State, SE Brazil. Width of view 18 mm. CPL.

Fig. 9.5.47 Detail of Fig. 9.5.46, showing strong undulose extinction in the large quartz crystal and fine-grained recrystallised grains in the tips of the fish-shaped structure. Sense of shear is sinistral. Width of view 6 mm. CPL.

9.6 Folds in Mylonites

Asymmetric folds (Fig. 9.6.1), often intrafolial, are relatively common in mylonites and their vergence can be used as a shear sense indicator if observed in sections parallel to the aggregate lineation (stretching lineation) and orthogonal to the mylonitic foliation, and if the fold axes are at a high angle to the stretching lineation. They tend to be tight to isoclinal with tapering hinges and are thought to develop during mylonitisation. However, some may be older, modified by the mylonitisation.

Irregularities in the flow pattern of a shear zone may produce folds at any stage of mylonite development. The older folds would be sheared out by subsequent mylonitisation and the younger folds would show a folded mylonitic fabric. In the sequence of photomicrographs shown below this relative order, from old to young with respect to the main mylonitisation, is followed. Lattice orientation of quartz tends to be strong in mylonites and may overprint old folds, whereas late folds will tend to show a folded fabric. This last situation is shown in Fig. 9.6.21, where the two limbs of a fold show different optical orientation.

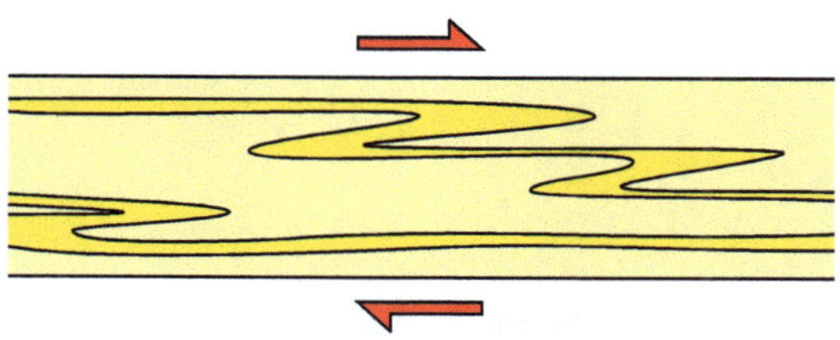

Fig. 9.6.1 Common shape of asymmetric folds in mylonites

Fig. 9.6.2 Ultramylonite with porphyroclasts of K-feldspar, derived from an orthogneiss. Several asymmetric folds indicate dextral sense of shear. The delta-type flow pattern around the large porphyroclast below the center also indicates dextral shear because of its stair stepping. A third indicator of dextral shear is the oblique fabric in the quartz-rich part in the lower left quadrant (Fig. 9.6.3). The strongly deformed nature of this quartz reflects relatively low temperature during and after mylonitisation. Roraima, northern Brazil. Width of view 12 mm. PPL.

Fig. 9.6.3 As Fig. 9.6.2. Width of view 12 mm. CPL.

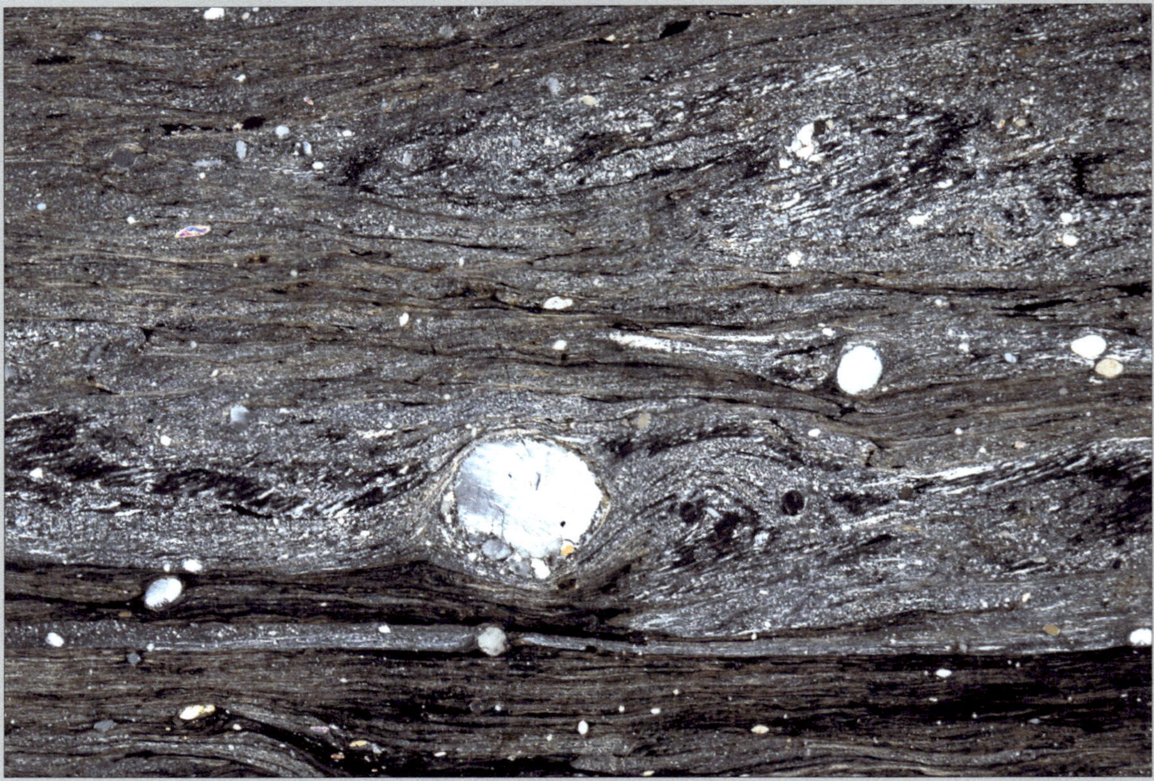

Fig. 9.6.4 Ultramylonite derived from granite, with porphyroclasts of K-feldspar. The upper part is very fine-grained and contains few porphyroclasts. Asymmetric folds in the central part indicate dextral shear. Notice the flow pattern around the relatively large porphyroclast below the center, also indicating dextral shear by its stair stepping. Low temperature shear zone from the Saint-Barthélemy Massif, Pyrenees, France.
Width of view 10 mm. PPL.

Fig. 9.6.5 Ultramylonite from the same shear zone as Fig. 9.6.4, showing asymmetric folds, indicating dextral sense of shear. Width of view 10 mm. PPL.

Fig. 9.6.6 Ultramylonite from the same shear zone as Fig. 9.6.4. Few porphyroclasts survived in this section. Asymmetric folds indicate dextral sense of shear. The strongly deformed quartz, visible in Fig. 9.6.7, indicates low metamorphic grade during mylonitization. The strong colour contrast between dark and clear layers is due to the presence of extremely fine-grained biotite in the dark layers. Width of view 15 mm. PPL.

Fig. 9.6.7 As Fig. 9.6.6. Width of view 15 mm. CPL.

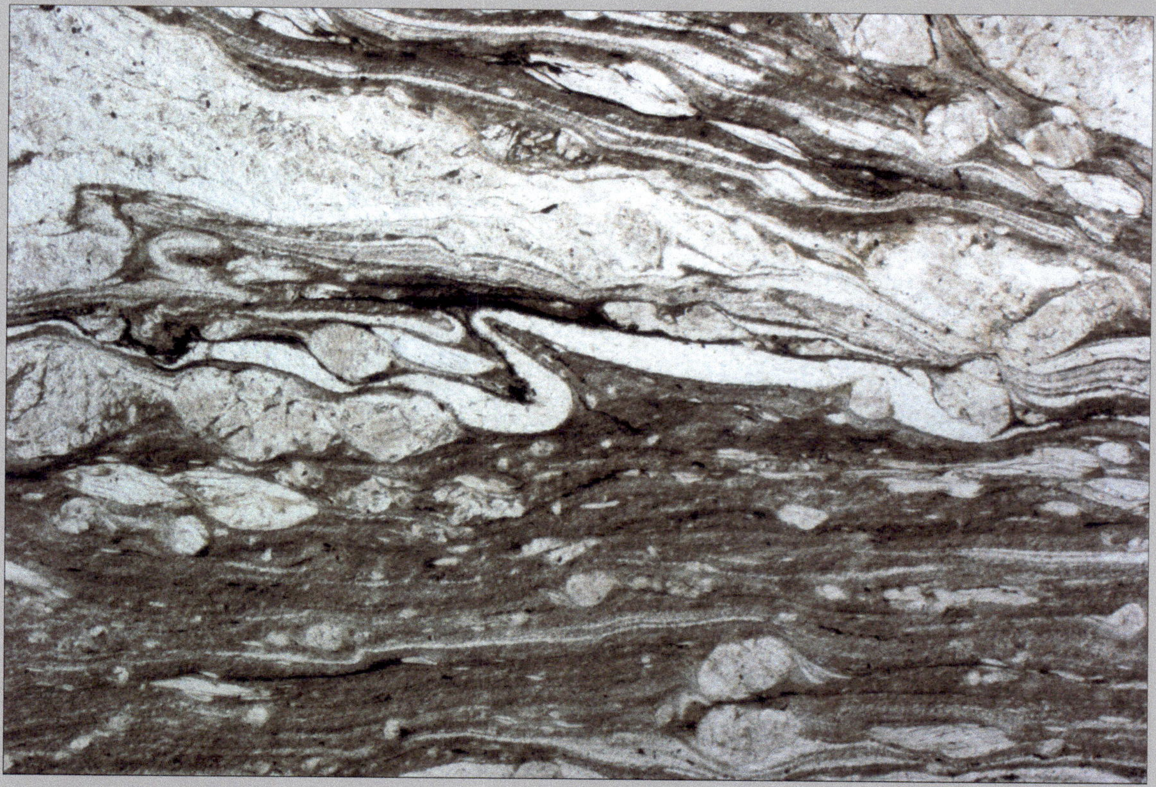

Fig. 9.6.8 Mylonite derived from granite with K-feldspar and plagioclase porphyroclasts. Asymmetric folds in the center and oblique mica fish in the upper part indicate sinistral sense of shear. Low-grade mylonite from a major shear zone in Pernambuco, NE Brazil. Width of view 10 mm. PPL.

Fig. 9.6.9 Mylonite with very fine-grained matrix derived from an amphibolite with porphyroclasts of plagioclase and hornblende. Asymmetric folds in the central and lower parts of the photo indicate sinistral sense of shear. Fig. 9.6.10 shows core-mantle structures of the plagioclase porphyroclasts with undulose cores surrounded by very fine-grained recrystallised plagioclase. The hornblende grains do not show this because they need higher temperature to recrystallise. Low-grade mylonite. Mt. West, Western Australia. Width of view 6 mm. PPL.

Fig. 9.6.10 As Fig. 9.6.9. Width of view 6 mm. CPL.

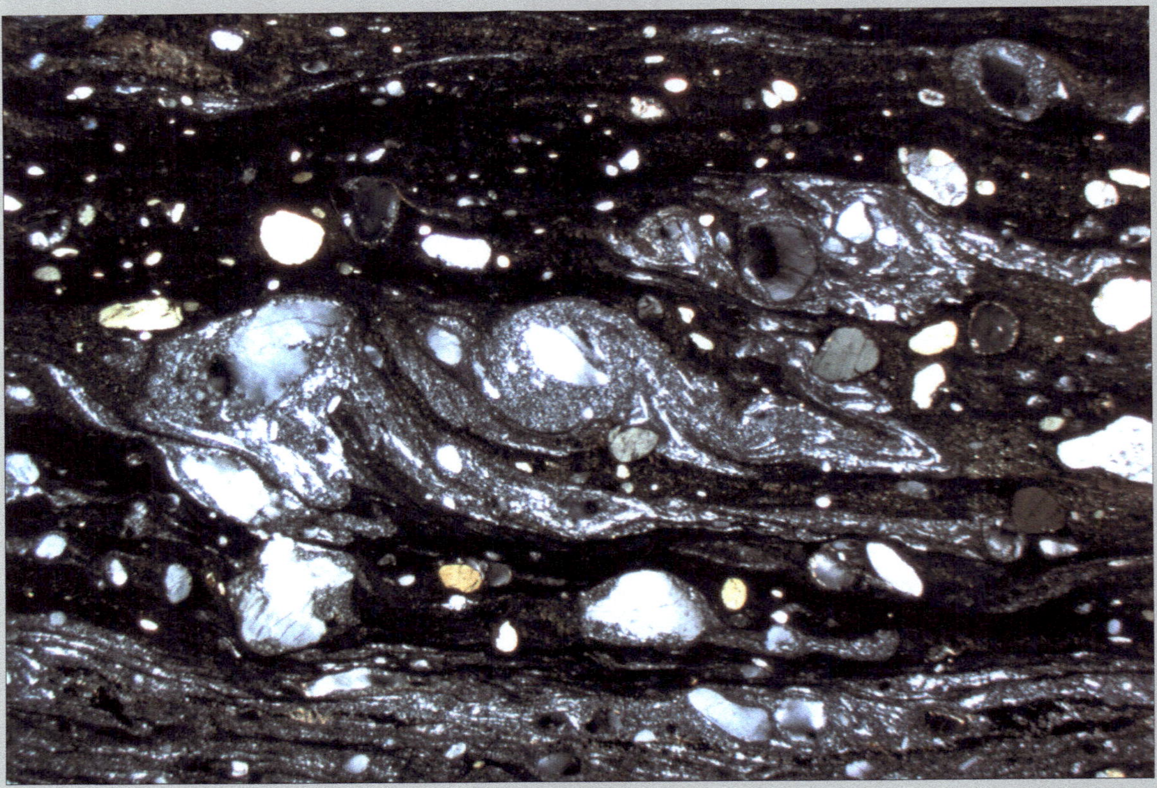

Fig. 9.6.11 Mylonite from a low-grade shear zone, derived from a paragneiss. The porphyroclasts are feldspar and muscovite. Dark bands contain very fine-grained biotite. Asymmetric folds, C/S fabric (lower left) and mica fish indicate sinistral shear. South Island, New Zealand. Width of view 22 mm. PPL.

Fig. 9.6.12 As Fig. 9.6.11. Width of view 22 mm. CPL.

Fig. 9.6.13 Ultramylonite (lower part) and mylonite (upper part), derived from a granitic rock. Isoclinal asymmetric folds, verging to the left, indicate sinistral sense of shear. The extremely fine-grained matrix and the partial recrystallisation of quartz (Fig. 9.6.14) indicate low-grade metamorphic conditions. Western Australia. Width of view 5 mm. PPL.

Fig. 9.6.14 As Fig. 9.6.13. Width of view 5 mm. CPL.

Fig. 9.6.15 This case is transitional between "folded mylonite", where the main mylonitic foliation is folded without much axial plane foliation and "asymmetric folds in mylonites", where compositional banding is tightly to isoclinaly folded with the main mylonitic foliation being axial planar to the folds. Notice that the difference between these two groups can be, in some cases, a matter of scale. Leiden collection . Width of view 28 mm. PPL.

Fig. 9.6.16 As Fig. 9.6.15. Width of view 28 mm. CPL.

Fig. 9.6.17 Folded mylonite of low- to medium-grade, derived from paragneiss, containing biotite, muscovite and garnet. The folded mylonitic foliation is approximately vertical and the trace of the axial plane of the folds is subhorizontal. Field evidence indicates that the fold was formed during mylonitisation. Marsfjällen, Västerbotten, Sweden. Width of view 20 mm. PPL.

Fig. 9.6.18 As Fig. 9.6.17. Width of view 20 mm. CPL.

Fig. 9.6.19 Folded mylonite from the same locality as Fig. 9.6.17. Width of view 20 mm. PPL.

Fig. 9.6.20 As Fig. 9.6.19. Width of view 20 mm. CPL.

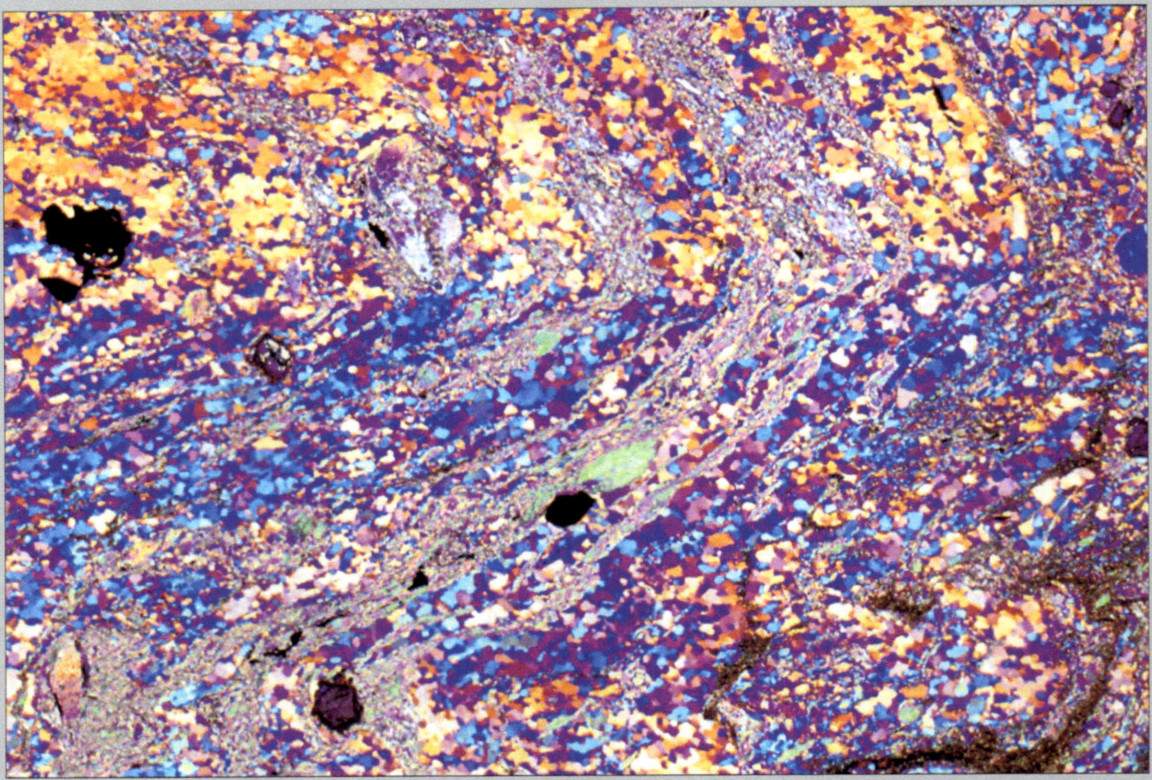

Fig. 9.6.21 Detail of a fold similar to that in Fig. 9.6.18, but with CPL and gypsum plate to show that the quartz lattice is folded. In the lower fold limb most quartz grains are blue whereas in the upper limb most are yellow as a consequence of their different optical orientation due to the folding. Marsfjällen, Västerbotten, Sweden. Width of view 15 mm. CPL with gypsum plate.

9.7 Other Shear Sense Indicators

In this section examples are shown of domino-type fragmented porphyroclasts (Fig. 9.7.1), shear-band type fragmented porphyroclasts (Fig. 9.7.5), quarter folds (Fig. 9.7.9), asymmetric strain shadows (Fig. 9.7.13) and domino-type boudinage (Passchier & Trouw 2005, their Fig. 5.41). The first group is characterised by antithetic movement along internal faults that make an angle of more than 45° with the mylonitic foliation. The second group includes porphyroclasts with internal faults at less than 45° to the mylonitic foliation and synthetic movement. The name quarter folds refers to their arrangement around porphyroclasts in opposite quarters. For example, in dextral shear they may develop in the upper right and lower left quarters. In a similar way as fragmented porphyroclasts, boudins can also be subdivided into domino-type and shear-band type (Passchier & Trouw 2005, their Fig. 5.41).

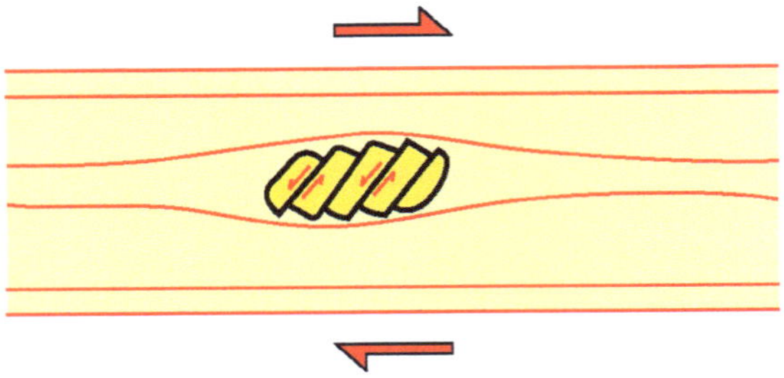

Fig. 9.7.1 Domino-type fragmented porphyroclast. Notice that the movement along the faults in the porphyroclast is antithetic to the sense of shear in the matrix and that the angle between these faults and the mylonitic foliation is usually larger than 45°.

Fig. 9.7.2 Domino-type fragmented porphyroclast of K-feldspar in a matrix composed of quartz, feldspar, mica and fragmented tourmaline. The rock is a protomylonite derived from a granite. Notice that the angle between the fault planes that separate the feldspar fragments and the mylonitic foliation exceeds 45° to the right of the large porphyroclast; below this porphyroclast the angle is smaller due to the compressional effect of the porphyroclast. Sense of shear between the fragments is sinistral, but the general sense of shear in the rock is dextral. This structure is also referred to as book-shelf sliding. Taxaquara, Sao Paulo State, SE Brazil. Width of view 3 mm. CPL.

Fig. 9.7.3 This low-grade mylonite derived from a granite shows a broken porphyroclast of feldspar with apparent dextral dislocation along the microfault that separates the two fragments. Since the fault makes a large angle with the mylonitic foliation the structure is interpreted as a domino-type fragmented porphyroclast, indicative of sinistral sense of shear in the mylonite, confirmed by several mica fish and some asymmetric minor folds. Pernambuco, NE Brazil. Width of view 4 mm. PPL.

Fig. 9.7.4 Mylonitised paragneiss with feldspar porphyroclasts and mica fish in a matrix mainly composed of quartz and mica. The central porphyroclast of plagioclase is broken into three pieces with dextral offset between them. However, the sense of shear in the mylonite is sinistral as indicated by several mica fish and stair stepping across the porphyroclast. The fragmented porphyroclast is therefore interpreted as domino-type. This example illustrates that the interpretation of shear sense from fragmented porphyroclasts is not always unequivocal. Pernambuco, NE Brazil. Width of view 2 mm. CPL.

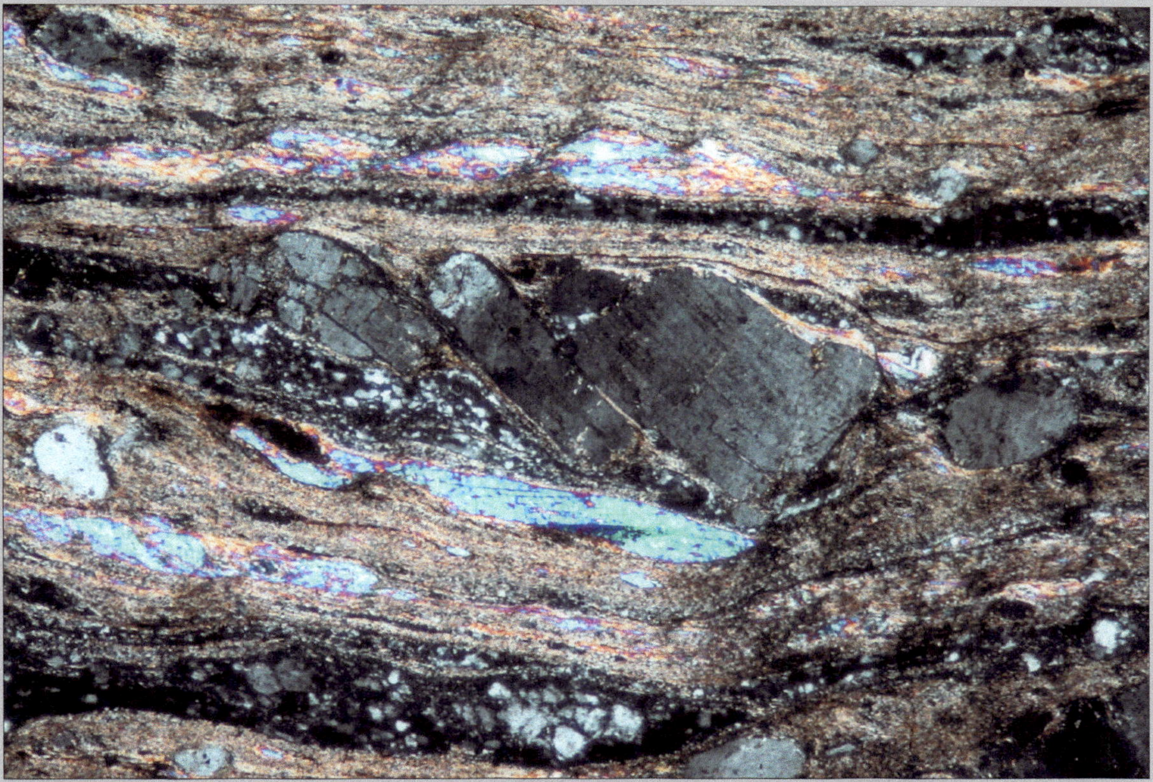

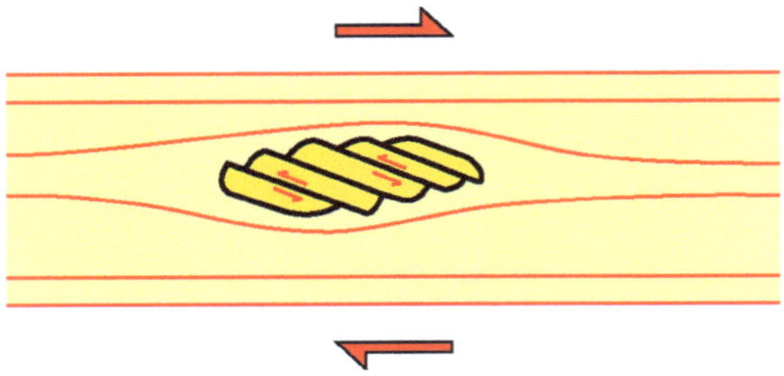

Fig. 9.7.5 Shear-band type fragmented porphyroclast. Notice that the movement along the faults in the porphyroclast is synthetic with the sense of shear in the surrounding mylonite and that the angle between the microfaults and the mylonitic foliation is smaller than 45°.

Fig. 9.7.6 Ultramylonite derived from a granitic rock showing a shear band type fragmented porphyroclast of K-feldspar embedded in a very fine-grained matrix. Sense of shear between the two fragments is dextral, syntaxial to the sense of shear in the mylonite. The metamorphic circumstances during mylonitisation were probably low grade, as indicated by the brittle behaviour of feldspar and the very fine-grained matrix. Australia. Width of view 5 mm. CPL.

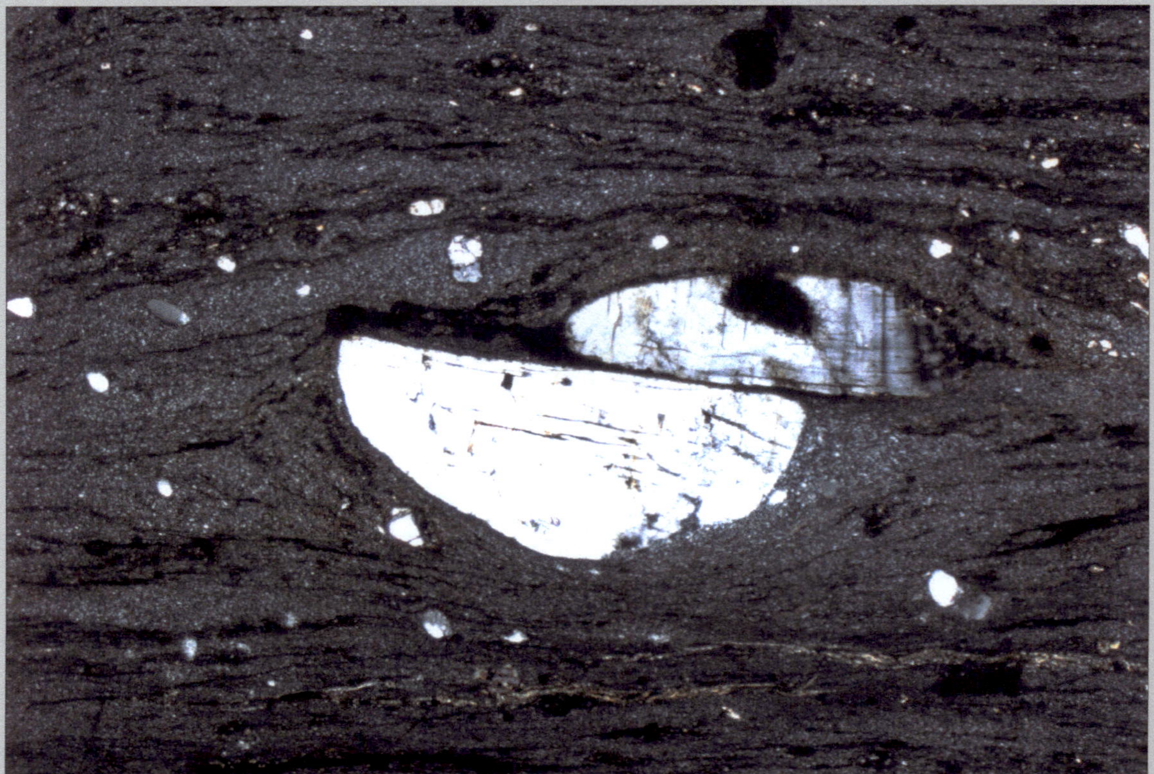

Fig. 9.7.7 Shear band type porphyroclast with sinistral sense of shear. The rock is a mylonite derived from a granite and the metamorphic conditions were low-grade, indicated by the partially recrystallised quartz (e.g. in a light band close to the bottom and at upper right). The sense of shear is confirmed by the stair stepping of the mylonitic foliation around the central porphyroclast and an incipient C/S fabric. Pernambuco, NE Brazil. Width of view 5 mm. CPL.

Fig. 9.7.8 Medium-grade mylonite with porphyroclasts of garnet and kyanite in a biotite-rich matrix. The parent rock was probably a garnet-biotite-kyanite schist (possibly a restite or melanosome from a migmatite). The garnet in the center is broken in two halves with an apparent dextral offset. Since the angle between the fault plane and the mylonitic foliation is less than 45°, the structure is interpreted as a shear-band type fragmented porphyroclast (however, the angle is very close to 45 degrees and the interpretation is therefore not unequivocal). Hence, the sense of shear in the rock was probably dextral as well, but the lack of other shear-sense indicators to confirm this makes the conclusion uncertain. Marsfjällen, Västerbotten, Sweden. Width of view 3 mm. PPL.

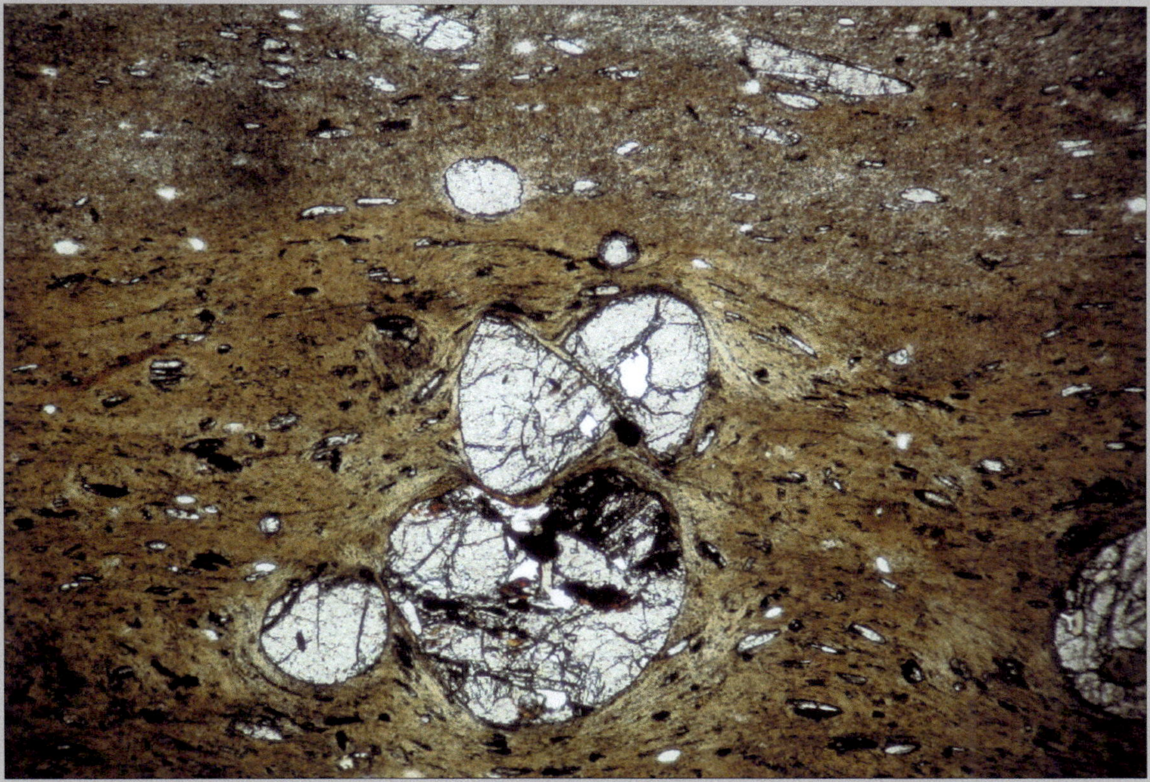

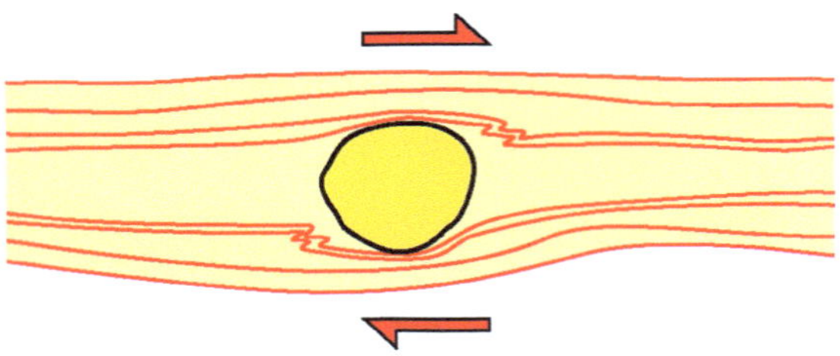

Fig. 9.7.9 Quarter folds. Notice that the folds in dextral shear only develop in the upper right and lower left quadrants.

Fig. 9.7.10 This mylonite derived from a granite contains a large porphyroclast of feldspar in the center. At the upper right side of this porphyroclast, some minor folds verging to the right define the structure as quarter folds, indicative of dextral shear sense. Other structures in the matrix, like mica fish and stair stepping across minor porphyroclasts confirm this sense of shear. Pernambuco, NE Brazil. Width of view 4 mm. PPL.

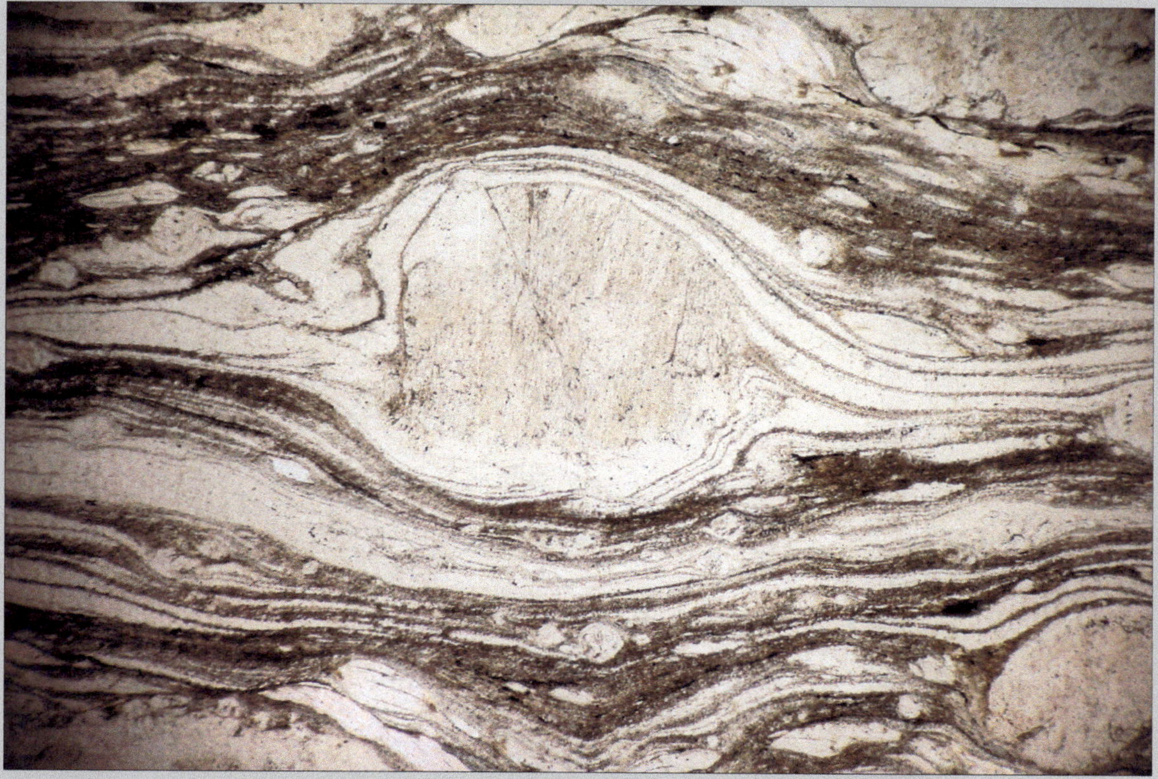

Fig. 9.7.11 Example of quarter folds around a feldspar porphyroclast from the same locality as Fig. 9.7.10. Notice the folds both in the upper left and the lower right quadrants, indicating sinistral sense of shear. Width of view 3 mm. PPL.

Fig. 9.7.12 Mylonite showing a porphyroclast in the center with quarter folds in the upper left and lower right quadrants, indicating sinistral sense of shear. Minor asymmetric folds and mica fish in the matrix reforce this interpretation. Pernambuco, NE Brazil. Width of view 3 mm. PPL.

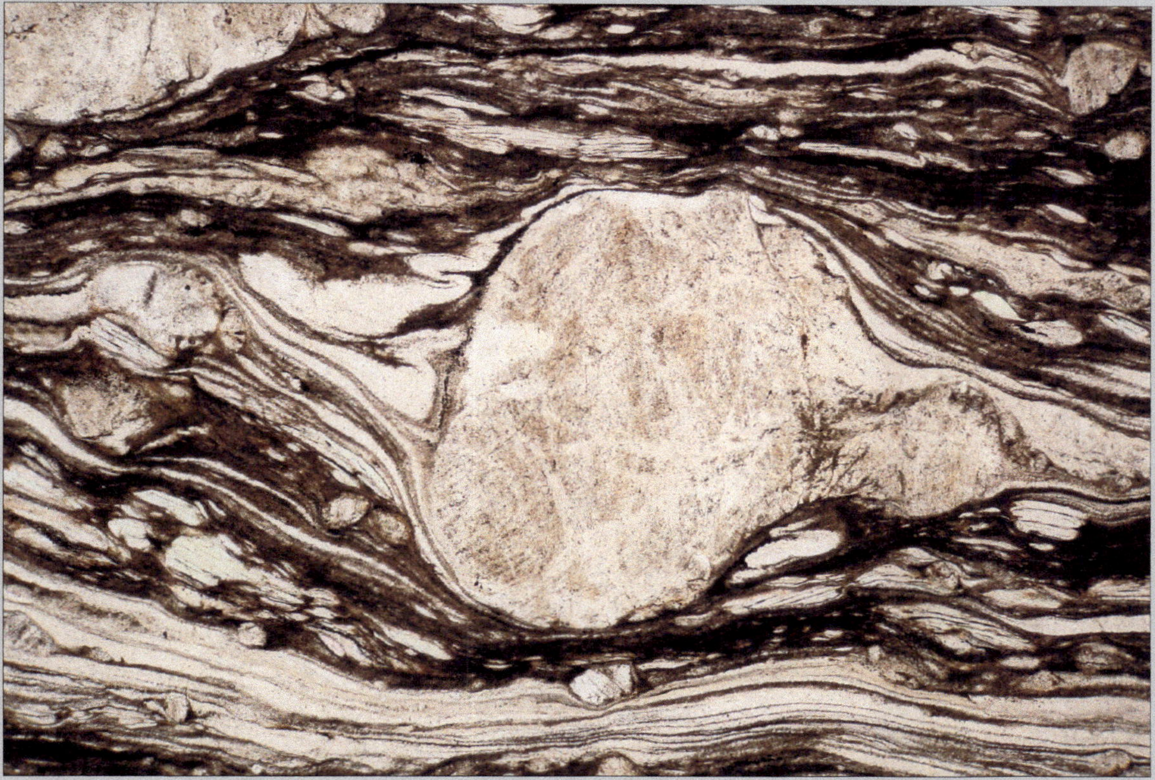

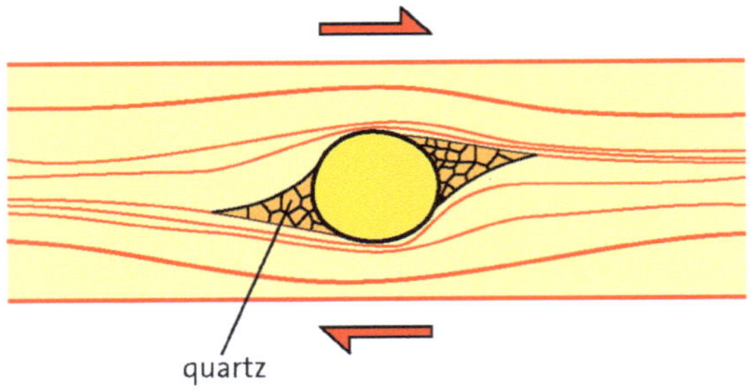

Fig. 9.7.13 Idealised model of asymmetric strain shadows.

Fig. 9.7.14 Porphyroclast of K-feldspar embedded in low-grade mylonite. The mylonitic foliation deviates strongly around the porphyroclast and shows accentuated stair stepping indicative for dextral sense of shear. Minor recrystallisation along the rim of the porphyroclast into very fine-grained new grains is just visible at this magnification (Fig. 9.7.15). The porphyroclast cannot be classified as a clear delta or sigma structure because it did not develop clear wings and is therefore better described as a porphyroclast with asymmetric strain shadows. Roraima, northern Brazil. Width of view 5 mm. PPL.

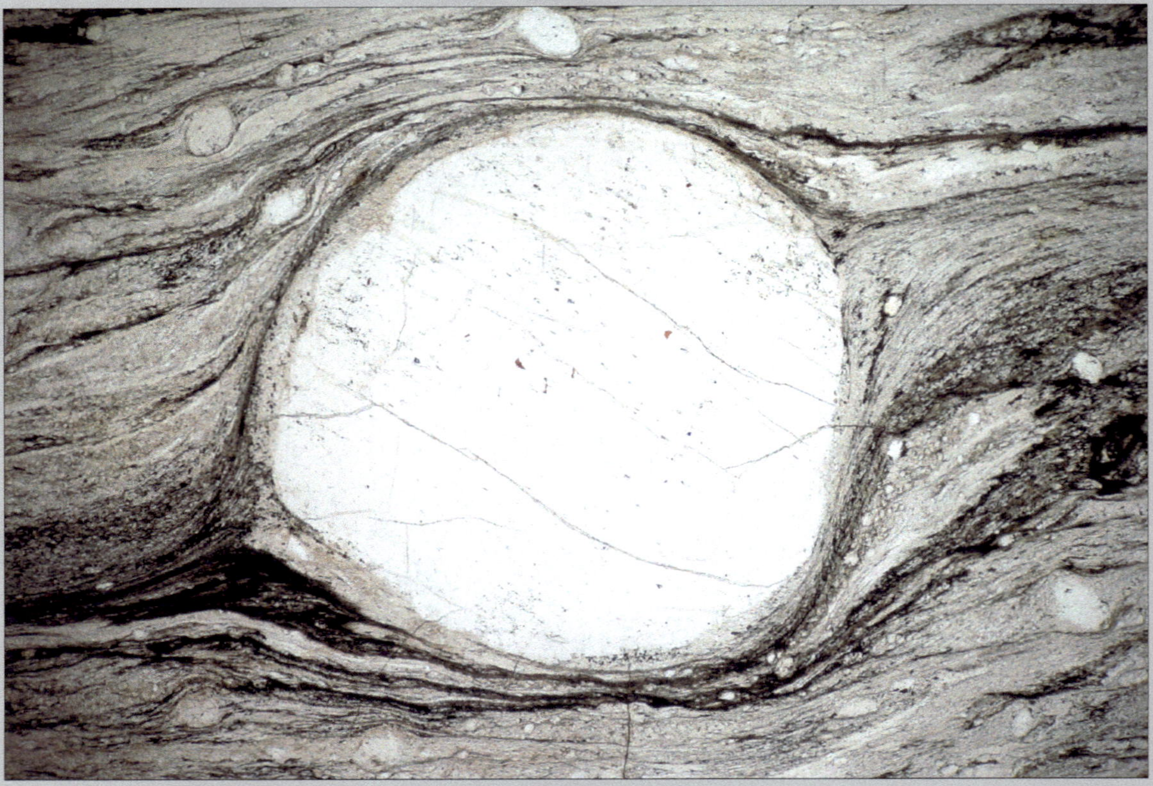

Fig. 9.7.15 As Fig. 9.7.14. Width of view 5 mm. CPL.

Fig. 9.7.16 Low-grade mylonite derived from a sillimanite bearing quartzite, showing sillimanite porphyroclasts in a matrix of strongly deformed quartz with minor recrystallisation by bulging. The sillimanite porphyroclasts are boudinaged and rotated in a dextral sense, whereas apparent offset between the boudins is sinistral. The structure is interpreted to show domino boudins related to dextral shear, confirmed by other shear sense indicators visible in the thin section but not in this photomicrograph. Saint Barthélemy Massif, French Pyrenees. Width of view 4 mm. CPL.

9.8 Superposed Sense of Shear

Shear zones are commonly reactivated at various stages of their evolution and this reactivation may include a change in the sense of shear. In general the new sense of shear will tend to overprint the old one beyond recognition, but in several cases it is possible to recognise an old sense of shear overprinted by a later one. An example of this situation is shown in this section.

Fig. 9.8.1 Mylonite derived from an amphibolitic rock with porphyroclasts of plagioclase and hornblende. The structure shows two large porphyroclasts in the center with a well developed core-and-mantle structure showing relatively large recrystallised feldspars surrounding a core with undulose extinction (Fig. 9.8.2). Their fish-like shape and stair stepping to the right is indicative for dextral sense of shear. Western Australia. Width of view 11 mm. PPL.

Fig. 9.8.2 As Fig. 9.8.1. Width of view 11 mm. CPL.

Fig. 9.8.3 Detail of Fig. 9.8.2 to show the recrystallisation around the core of the central porphyroclasts, indicative of medium-grade metamorphic conditions. Width of view 5 mm. CPL.

Fig. 9.8.4 Close-up of the upper right-hand part of Fig. 9.8.3 showing several quartz veins at left, with deformed and recrystallised quartz. Oblique foliation and an asymmetric fold in quartz are indicative of sinistral shear under low-grade metamorphic conditions, apparently superposed on the older dextral shear. Width of view 2 mm. CPL.

Chapter 10 | Crystal-Plastic Deformation, Recovery and Recrystallisation of Quartz

10 Crystal-Plastic Deformation, Recovery and Recrystallisation of Quartz

As stated in the introduction, this chapter is included because of the special importance of quartz to estimate metamorphic conditions during and after mylonitisation. The theory behind crystal-plastic deformation is treated elsewhere (e.g. Passchier & Trouw 2005). The main optical expression of crystal-plastic deformation is smooth, non-patchy undulose extinction. Elongated grains with such undulose extinction, sometimes accompanied by deformation lamellae, are indicative for low-temperature deformation. At slightly higher temperatures recovery produces subgrains and recrystallisation tends to substitute the old deformed grains by small new ones. Three types of recrystallisation can be distinguished (Fig. 10.1):

- bulging recrystallisation at relatively low temperature; low temperature grain boundary migration causes small-amplitude lobate grain contacts that eventually separate small grains by a process of "necking";
- subgrain rotation recrystallisation at low to medium temperature; this type of recrystallisation produces new grains by the progressive rotation of subgrains;
- high-temperature grain boundary migration; strictly speaking, no new grains are produced by this mechanism, but growth by large-amplitude grain boundary migration produces large grains with irregular shapes and often with inclusions of other minerals.

If temperature is sufficiently high after subgrain rotation recrystallisation, adjustment of grain boundaries under static conditions tends to result in a polygonal granoblastic fabric, typical for medium-grade metamorphic conditions.

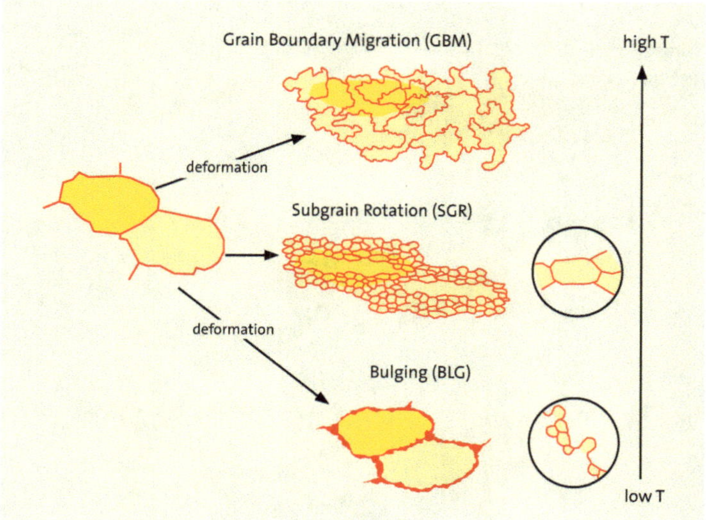

Fig. 10.1 The three main types of dynamic recrystallisation in a polycrystalline aggregate. The substance of one of two large grains that recrystallise is indicated in bright yellow, before and during recrystallisation.

Fig. 10.2 Low-grade metawacke with detrital quartz grains in a matrix of muscovite and fine-grained quartz. The horizontal secondary foliation is deflected around the more resistant quartz grains that show undulose extinction of sweeping type due to crystal-plastic deformation. However, the approximately equidimensional shape of the quartz grains shows that they are little deformed. This rock is not a mylonite which is clear from study in outcrop where an angle between bedding and cleavage can be observed. However, at first sight, the structure is similar (see also Chapter 11). Quartz remained in its deformed state because of the low metamorphic grade. Southern Minas Gerais State, SE Brazil. Width of view 5 mm. CPL.

Fig. 10.3 Low-grade metawacke with detrital quartz grains in a matrix of muscovite and fine-grained quartz. Deformation in this rock was more intensive than in the rock shown in Fig. 10.2, as can be deduced from the more elongated shape of the quartz grains. Quartz shows sweeping undulose extinction, indicative of low-grade metamorphic conditions. Southern Minas Gerais State, SE Brazil. Width of view 12 mm. CPL.

Fig. 10.4 Low-grade metawacke with strongly elongated, mainly monocrystalline quartz grains in a matrix of muscovite and fine-grained quartz. The quartz grains are strongly deformed by crystal-plastic deformation, forming ribbons with sweeping undulose extinction. This rock is transitional to a mylonite. Some quartz grains are more deformed than others, probably due to different lattice orientation. Western Alps. Width of view 12 mm. CPL.

Fig. 10.5 Quartzite deformed under low-grade metamorphic conditions. Most quartz grains show undulose extinction, some of the sweeping type and others more patchy. The contacts between the quartz grains tend to become lobate by grain-boundary migration. Bahia, NE Brazil. Width of view 8 mm. CPL.

Fig. 10.6 Quartzite deformed under low-grade metamorphic conditions. Note the elongated shape of most grains, defining a horizontal secondary foliation. Most quartz grains show sweeping undulose extinction. Note the lobate contacts that locally isolate small new grains by bulging recrystallisation. Pernambuco, NE Brazil. Width of view 10 mm. CPL.

Fig. 10.7 Low-grade deformed quartzite similar to the one shown in Fig. 10.6. Note the subvertical cataclastic shear band, right of the center with strongly reduced grain size due to cataclasis. Guapé, southern Minas Gerais State. Width of view 13 mm. CPL.

Fig. 10.8 Low-grade quartzite with sweeping undulose extinction and lobate contacts due to grain boundary migration. At some localities (e.g. extreme right) new grains are being produced by bulging recrystallisation. Bahia, NE Brazil. Width of view 5 mm. CPL.

Fig. 10.9 Low-grade quartzite with deformed quartz showing sweeping undulose extinction, lobate contacts due to grain boundary migration and deformation lamellae (e.g. upper right–lower left) in the central grain. Bahia, NE Brazil. Width of view 3 mm. CPL.

Fig. 10.11 Impure quartzite with opaque minerals (upper left) deformed under low-grade metamorphic conditions. The central quartz grain is strongly deformed by crystal-plastic deformation resulting in chevron folds within the quartz with subvertical axial plane. Subhorizontal deformation lamellae are just visible in the central and right part of the photomicrograph. Lobate contacts and recrystallisation by bulging can be observed at lower left. Bahia, NE Brazil. Width of view 3 mm. CPL.

Fig. 10.12 Low-grade micaceous quartzite with two orthogonal metamorphic foliations. The horizontal foliation defined by the preferred orientation of muscovite is older than the vertical foliation defined by the elongated shape of quartz crystals. This is evident from small folds in the muscovites with vertical trace of the axial plane. Many quartz grains show undulose extinction indicating that the vertical shape fabric of the quartz was induced by horizontal shortening under low-grade conditions. The photomicrograph demonstrates that crystal-plastic deformation of quartz is capable of inducing a shape fabric with relatively little deformation; in the case of stronger deformation, the mica foliation would not have survived in its horizontal position. Itumirim, southern Minas Gerais State, SE Brazil. Width of view 12 mm. CPL.

Fig. 10.13 Similar situation as shown in Fig. 10.12. Elongate chlorite crystals define an older foliation trending upper left to lower right. Elongate quartz grains trend upper right to lower left; they show undulose extinction and lobate contacts related to the last deformation experienced by this quartzitic rock. Some gentle folds in the chlorite foliation confirm the sequential relation between the two foliations. Itumirim, southern Minas Gerais State, SE Brazil. Width of view 2.5 mm. CPL.

Fig. 10.14 Low-grade quartzite with large grains deformed by crystal-plastic deformation. Most contacts are lobate. The undulose extinction shows some relatively sharp transitions in grey tones, defining elongated subgrains (upper left-lower right); the sweeping undulose extinction changes to patchy. Bahia, NE Brazil. Width of view 2 mm. CPL.

Fig. 10.15 Low-grade quartzite with lobate contacts due to grain boundary migration. At several sites (e.g. left of center) isolated small grains appear as a result of bulging recrystallisation. The central grain shows a transition between patchy and sweeping undulose extinction, closer to patchy. Bahia, NE Brazil. Width of view 2 mm. CPL.

Fig. 10.16 Quartz-rich lens in a mylonitised gneiss showing strongly lobate contacts due to grain boundary migration. Bulging recrystallisation is apparent in the central part of the photomicrograph. Marsfjällen, Västerbotten, Sweden. Width of view 1.25 mm. CPL.

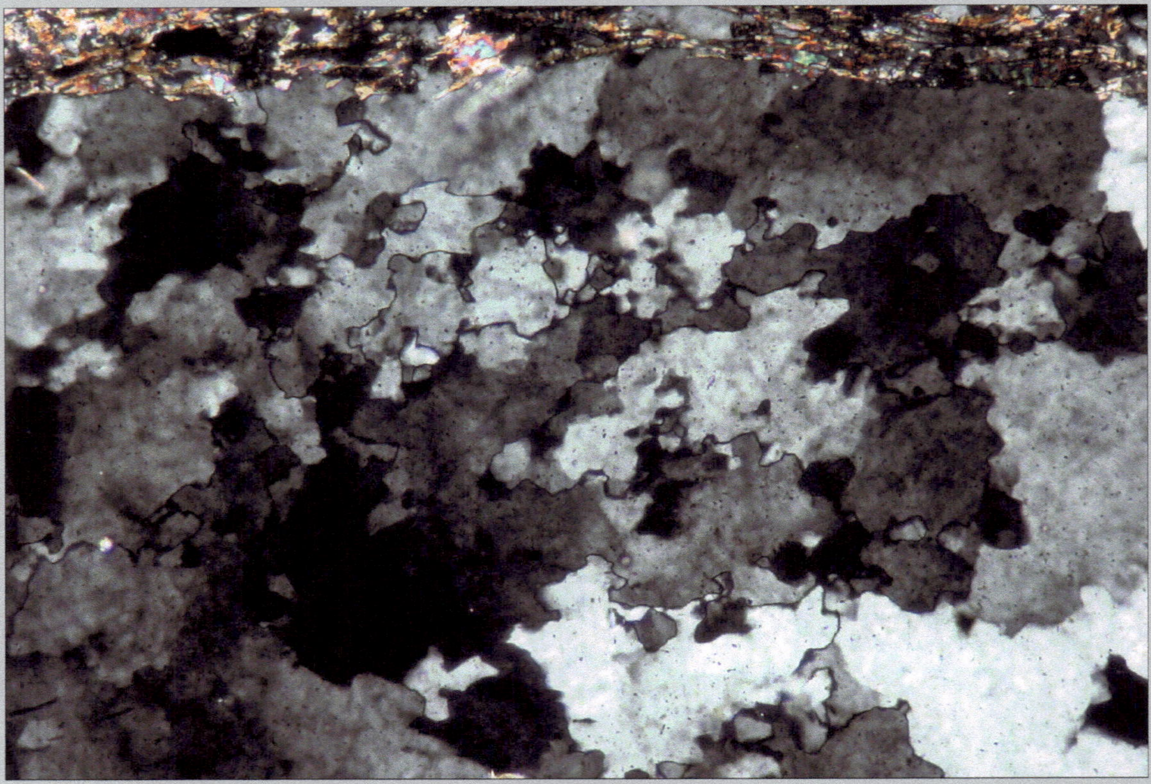

Fig. 10.17 Quartz-rich lens in a mylonitised gneiss with lobate contacts between quartz grains and incipient bulging recrystallisation (e.g. the white grains at lower right). Marsfjällen, Västerbotten, Sweden. Width of view 1.25 mm. CPL.

Fig. 10.18 Low-grade quartzite showing patchy undulose extinction and partial recrystallisation into about 20% new small grains. The mechanism of recrystallisation is probably both bulging (e.g. lower left) and subgrain rotation (e.g. central part). Bahia, NE Brazil. Width of view 1.8 mm. CPL.

Fig. 10.19 Low-grade quartzite with patchy undulose extinction in large old grains, defining subgrains. The transition to new grains with stronger contrasting grey tones is gradational, suggesting that subgrain rotation was the main mechanism of recrystallisation. Bahia, NE Brazil. Width of view 1.2 mm. CPL.

Fig. 10.20 Low-grade quartzite with subgrains within large deformed old grains, and small new grains appearing within a single old grain. The mechanism of recrystallisation was probably subgrain rotation since bulging occurs exclusively along lobate contacts of old grains. Bahia, NE Brazil. Width of view 3 mm. CPL.

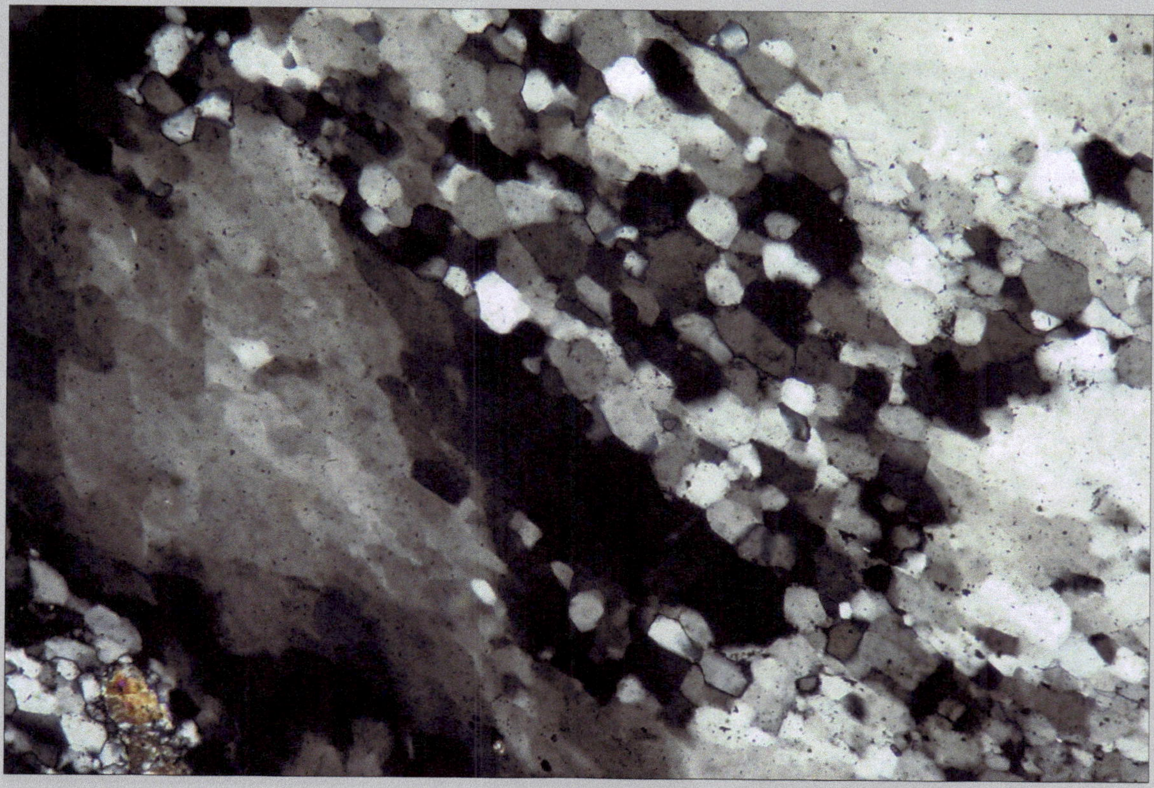

Fig. 10.21 Detail of the same thin section shown in Fig. 10.20. The whole photomicrograph shows an area inside one old deformed grain. At left the old grain is still recognizable by the patchy undulose extinction corresponding to subgrains. In the central part new grains, produced by subgrain rotation, predominate. At right the proportion is about 50% each. Bahia, NE Brazil. Width of view 1.2 mm. CPL.

Fig. 10.22 Partially recrystallised old quartz grain with gradual transition between subgrains and new grains. The mechanism of recrystallisation is subgrain rotation. Bahia, NE Brazil. Width of view 1.2 mm. CPL.

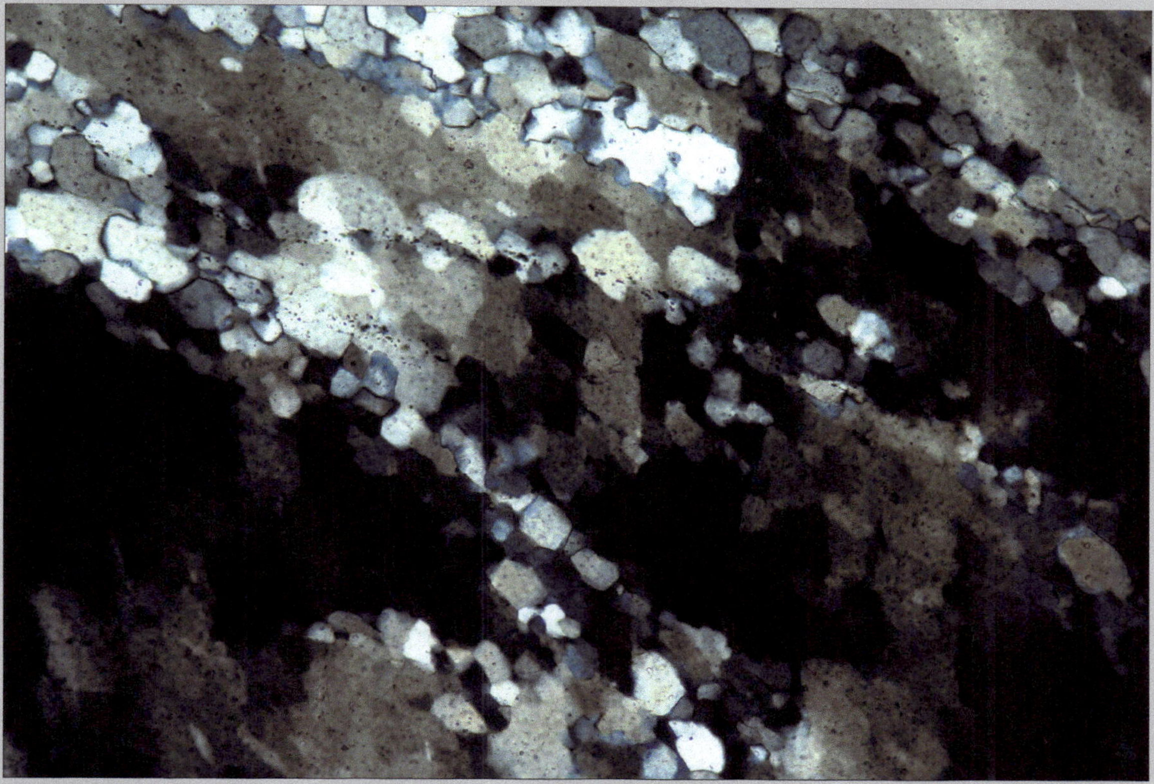

Fig. 10.23 Low-grade strongly deformed quartzite with about 25% recrystallisation. The elongated large old grains show strong sweeping undulose extinction. Recrystallisation was probably both by bulging and by subgrain rotation. Itumirim, southern Minas Gerais State, SE Brazil. Width of view 12 mm. CPL.

Fig. 10.24 Low-grade deformed quartzite showing about 40% recrystallisation. The large old grains exhibit undulose extinction. At this magnification it is difficult to distinguish between recrystallisation by bulging or by subgrain rotation. Conceição do Rio Verde, southern Minas Gerais State, SE Brazil. Width of view 10 mm. CPL.

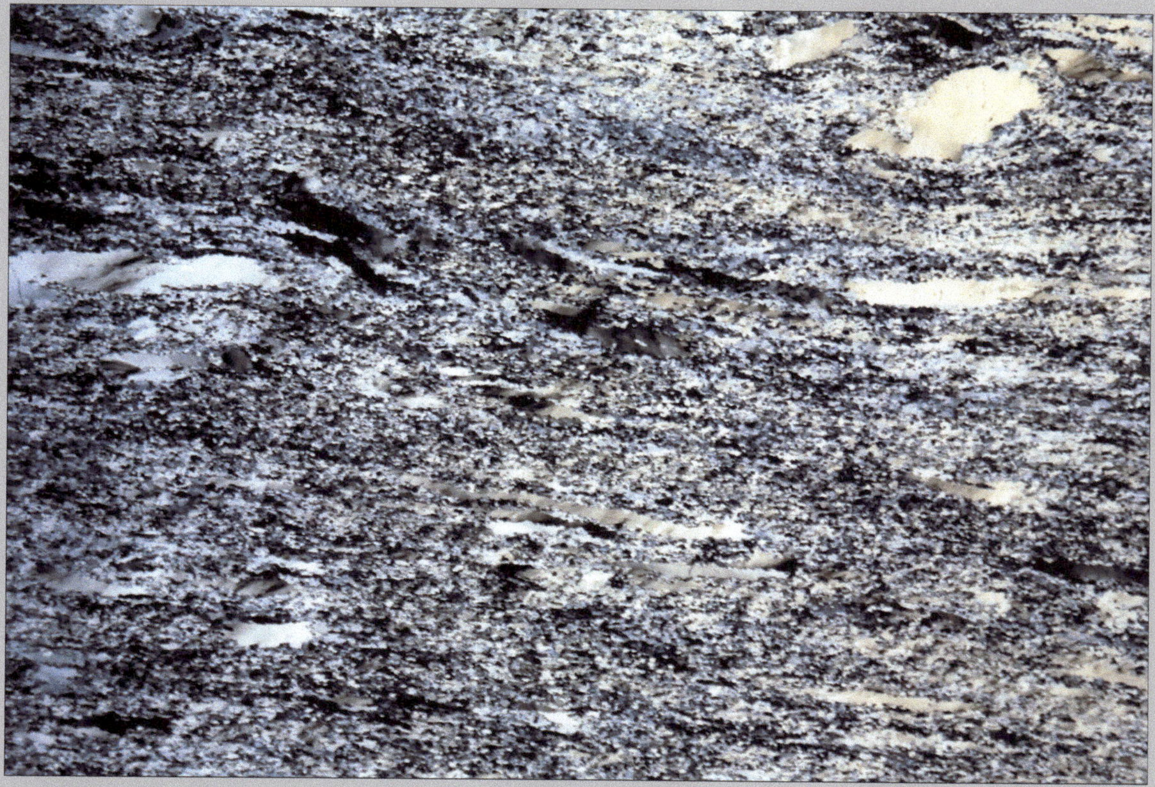

Fig. 10.25 Low-grade deformed quartzite with about 70% recrystallisation. Few old elongated grains with undulose extinction are recognizable. The predominant mechanism of recrystallisation was probably subgrain rotation, but at this magnification this is hard to see. Itumirim, southern Minas Gerais State, SE Brazil. Width of view 15 mm. CPL.

Fig. 10.26 Low-grade deformed quartzite with about 90% recrystallisation. Very few elongated old grains can be recognized (upper right). Itumirim, southern Minas Gerais State, SE Brazil. Width of view 15 mm. CPL.

Fig. 10.27 Fully recrystallised low- to medium-grade quartzite. Note the predominantly straight contacts that tend to make angles of about 120°. Light and dark clusters of grains reflect old recrystallised grains to some extend since the angle between the lattice of new and old grains is usually between 5 and 30°. Bahia, NE Brazil. Width of view 10 mm. CPL.

Fig. 10.28 Fully recrystallised low- to medium-grade quartzite. Note the predominantly equidimensional shape, the straight contacts and the tendency of these contacts to make angles of 120° at triple junctions. Bahia, NE Brazil. Width of view 5 mm. CPL.

Fig. 10.29 Fully recrystallised low- to medium-grade quartzite with relatively small equidimensional grains. The straight contacts tending to make angles of 120° are the result of static readjustment after deformation. Bahia, NE Brazil. Width of view 10 mm. CPL.

Fig. 10.30 Close-up of fully recrystallised quartz fabric with static adjustment of grain boundaries. Note the equidimensional shape of grains and the straight contacts that tend to make angles of 120° at triple junctions (e.g. upper left). Itumirim, southern Minas Gerais State, SE Brazil. Width of view 1.2 mm. CPL.

Fig. 10.31 Phyllite with quartz veins and folds with the trace of the axial surface trending ENE-WSW. The major quartz vein is folded but the quartz fabric is free of strain, demonstrating the complete recrystallisation and static readjustment after folding. Two deformation phases can be recognized in this thin section. The first one (D1) generated a slaty cleavage that is now trending upper left to lower right. The second phase (D2) produced the folds and a new spaced crenulation cleavage, trending upper right to lower left. The quartz veins must be pre-D2, because they are folded by D2. The well-recrystallised quartz in the veins shows that metamorphic conditions during and/or after D2 were sufficiently high for recrystallisation to proceed. Itumirim, southern Minas Gerais State, SE Brazil. Width of view 8 mm. CPL.

Fig. 10.32 Quartz fabric showing initiation of grain growth by (high-temperature) grain boundary migration. Several equidimensional grains with straight contacts at angles of 120° can still be recognized (e.g. below center), but other grains assume large irregular shape (e.g. above center) due to grain boundary migration. Santa Rosa Mylonite zone, California. Width of view 3 mm. CPL.

Fig. 10.33 Medium- to high-grade micaceous quartzite with large quartz grains containing several inclusions of muscovite. Note the irregular shape of the quartz grains due to grain boundary migration. At lower metamorphic grade quartz rarely contains inclusions. Morro do Cara de Cão, Rio de Janeiro, SE Brazil. Width of view 3 mm. CPL.

Fig. 10.34 Medium- to high-grade micaceous quartzite with microstructures typical for high-temperature grain boundary migration. Just left of the center is a "window" microstructure where two biotite grains (colored) impeded the advance of the light grey upper quartz grain into the dark one below. Hence, the movement direction of the migrating grain boundary was downwards. Further to the right is a "pinning" microstructure, where another biotite grain hampered the movement of a grain boundary in the same direction. Morro do Cara de Cão, Rio de Janeiro, SE Brazil. Width of view 1.2 mm. CPL.

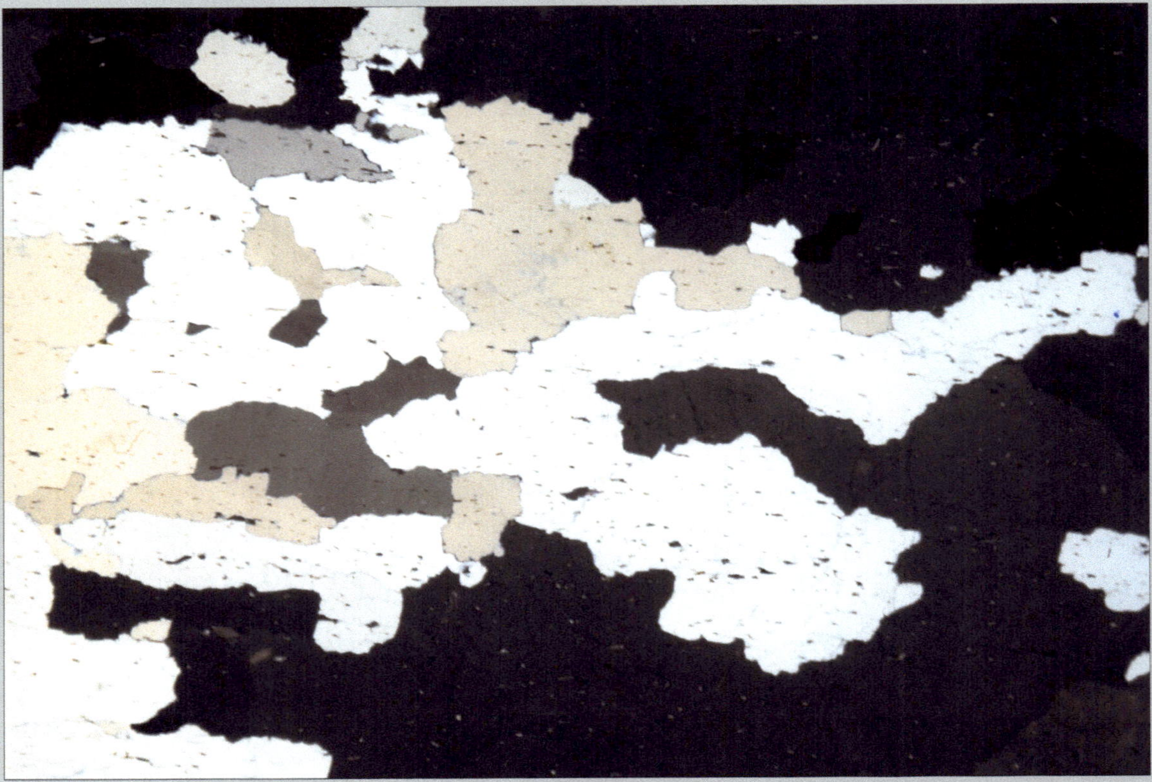

Fig. 10.35 Typical microstructure of high-grade micaceous quartzite. Quartz grains are large and very irregular in shape due to high-temperature grain boundary migration. Notice the large number of inclusions. Isolated parts with the same grey tone may represent parts of a single crystal that link in the third dimension, above or below the thin section. Juiz de Fora, southern Minas Gerais State, SE Brazil. Width of view 13 mm. CPL.

Fig. 10.36 High-grade micaceous quartzite with large quartz grains. The irregular grain boundaries and the presence of inclusions indicate high-temperature grain boundary migration as the mechanism for grain growth. The central grain shows undulose extinction indicating renewed deformation after grain growth.
Juiz de Fora, southern Minas Gerais State, SE Brazil. Width of view 3 mm. CPL.

Fig. 10.37 High-grade micaceous quartzite with large irregular shaped quartz grains due to high-temperature grain boundary migration. Renewed deformation at lower temperature induced undulose extinction with well-delineated subgrains of "chessboard" type. Note that the elongated shape of subgrains is controlled by the main shortening direction and the lattice orientation of the deforming crystal. Rio de Janeiro State, SE Brazil. Width of view 3 mm. CPL.

Fig. 10.38 High-grade sillimanite-bearing mylonitic quartzite, showing a fabric typical for grain growth by high-temperature grain boundary migration. The irregular shape of the quartz grains is controlled to some extend by sillimanite and biotite grains with a strong preferred orientation (horizontal). The undulose extinction reveals later deformation under lower temperature conditions. Maria da Fé, southern Minas Gerais State, SE Brazil. Width of view 18 mm. CPL.

Fig. 10.39 High-grade mylonite with quartz ribbons separated by fine-grained recrystallised K-feldspar. This photomicrograph is also shown in Fig. 6.1 as an example of a high-grade mylonite. It is probably derived from pegmatite since the rock is exclusively composed of K-feldspar and quartz. The peculiar shape of the quartz grains is due to extreme stretching followed by recrystallisation by high-temperature grain boundary migration. The grain boundaries could only advance along the stretched quartz grains since these are embedded in small recrystallised K-feldspar that would not permit lateral growth. Três Rios, Rio de Janeiro State, SE Brazil. Width of view 18 mm. CPL.

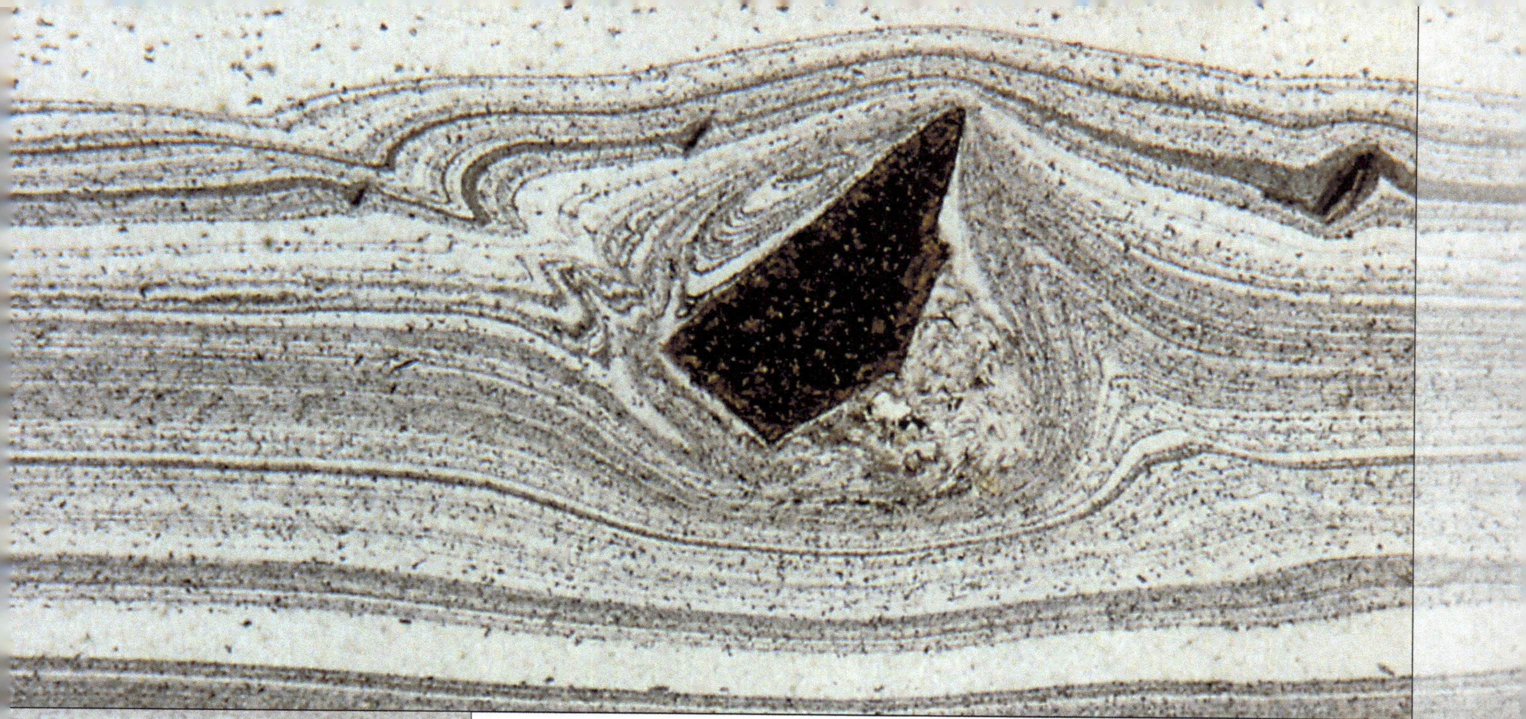

Chapter 11 | "False" Mylonites

11 "False" mylonites

Several little deformed sedimentary and volcanic rocks may contain structures that closely resemble mylonitic structures. Two examples are shown in this section. Low grade metasedimentary wackes containing detrital quartz grains in a fine-grained matrix may resemble mylonites. The quartz may show undulose extinction and resemble porphyroclasts in a mylonite (compare Fig. 10.2). However, in real mylonites quartz is the weaker mineral and tends to form the matrix rather then the porphyroclasts. An exeption to this are mylonitised ignimbrites or other volcanic rocks (e.g. Figs. 9.5.46 and 9.5.47; Passchier & Trouw 2005, their Figs. 3.9 and 3.10).

Another example is given by flow structures in lava. These can be differentiated from true mylonites by the fact that mylonites tend to contain rounded porphyroclasts whereas lavas usually contain angular phenocrysts.

Fig. 11.1 Low-grade metasediment with abundant clastic quartz grains in a mica-rich matrix. Ductile deformation has induced a horizontal cleavage. Deformation of the quartz grains by crystal-plastic processes resulted in undulose extinction. Low metamorphic grade has hampered recovery and recrystallisation. At first sight the rock looks like a mylonite with deformed "porphyroclasts" in a fine-grained matrix; however, quartz in real mylonites tends to form the matrix and usually does not survive as porphyroclasts (see also Fig. 10.2). Southern Minas Gerais State, SE Brazil. Width of view 12 mm. CPL.

Fig. 11.2 Low-grade metasediment with preserved bedding (upper right-lower left) at a small angle to slaty cleavage (horizontal). Deformed detrital quartz grains in abundant matrix resemble a mylonitic structure (Fig. 11.3). However, the preserved angle between bedding and slaty cleavage shows that the rock is not a mylonite. Southern Minas Gerais State, SE Brazil. Width of view 16 mm. PPL.

Fig. 11.3 As Fig. 11.2. Width of view 16 mm. CPL.

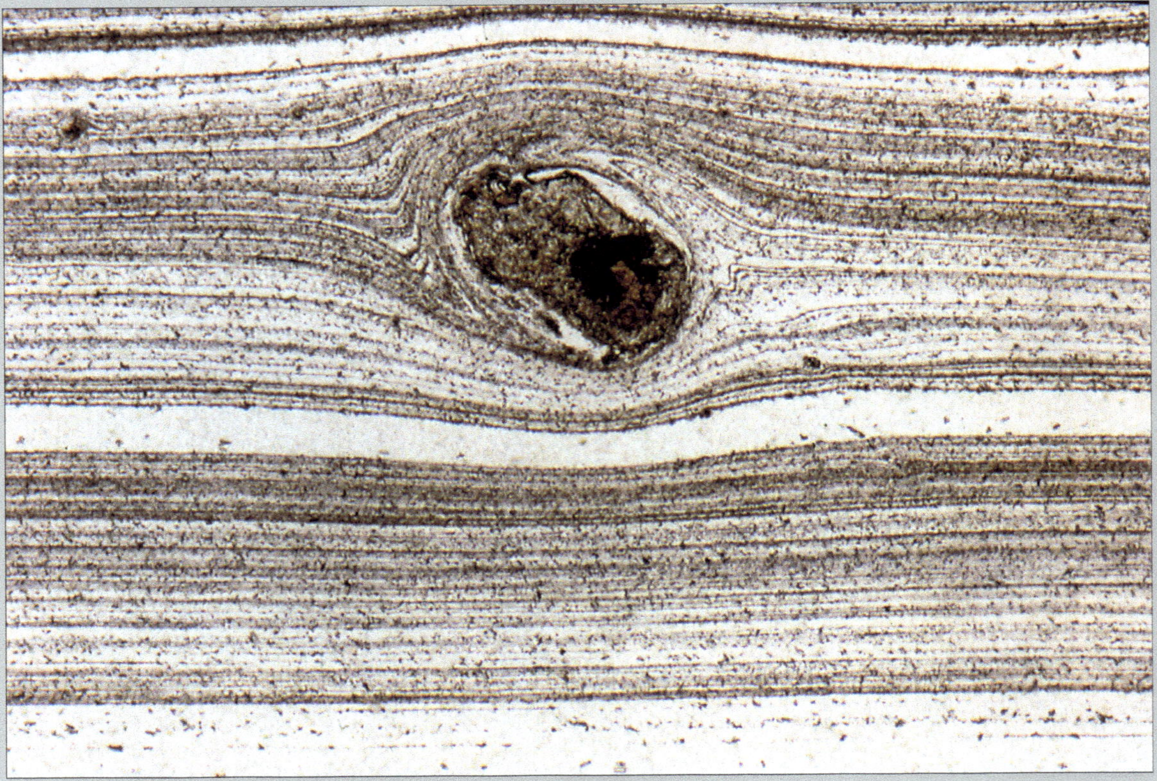

Fig. 11.4 Flow structure in lava with deviation and local folding around angular fragments. Sense of flow is sinistral. The structure looks similar to a mylonite but porphyroclasts in mylonites tend to be rounded because of a less extreme viscosity contrast. Width of view 5 mm. PPL. (thin section courtesy Ron Vernon).

Fig. 11.5 As Fig. 11.4. Note the complex flow pattern around the rigid clast, resembling quarter structure. Width of view 5 mm. PPL.

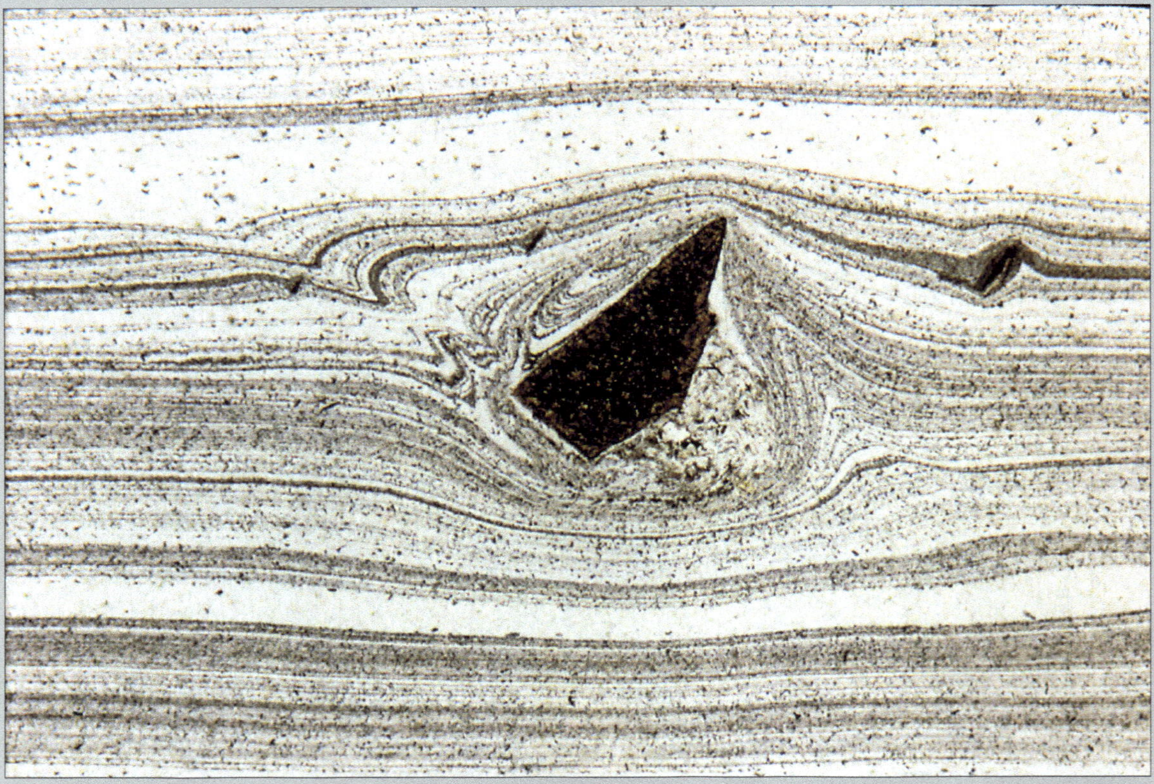

Fig. 11.6 As Fig. 11.4. Width of view 5 mm. PPL.

Fig. 11.7 Photomicrograph of the intrusive contact between a diabase dyke (at right) and amphibolite with horizontal schistosity (at left). The sharp contact and the almost opaque matrix of the dyke resemble pseudotachylyte, but the elongated plagioclase crystals leave no doubt about the igneous origin of this dyke. Buzios, Rio de Janeiro State, SE Brazil. Width of view 6 mm. PPL.

Fig. 11.8 As Fig. 11.7. Width of view 6 mm. CPL.

Fig. 11.9 Very fine-grained metasediment with isoclinal folds in the bedding and a well developed axial plane cleavage. The rock looks mylonitic but isoclinal folds in mylonites tend to have a more tapering shape in the hinges. Portugal. Width of view 16 mm. PPL.

Fig. 11.10 As Fig. 11.9. Width of view 16 mm. CPL.

Fig. 11.11 Deformed rhyolite. This rock resembles at first sight a mylonite by the presence of relatively large quartz grains (apparent porphyroclasts) in a fine-grained foliated matrix. The idiomorphic shape of the quartz (see also detail in Fig. 11.12) and the lack of undulose extinction reveal that the rock did not suffer strong deformation and that the contrast between large crystals in a fine-grained matrix is an original feature of this rock. This rock can be compared with Figs. 9.5.46 and 47 that were derived from a similar parent rock but with much more intensive deformation. Southern Chile. Width of view 16 mm. CPL.

Fig. 11.12 Detail of Fig. 11.11. The hypidiomorphic shape of this quartz crystal with an embayment is characteristic for magmatic growth. Note the lack of undulose extinction, indicating that the deformation was not very strong. Width of view 2.5 mm. CPL.

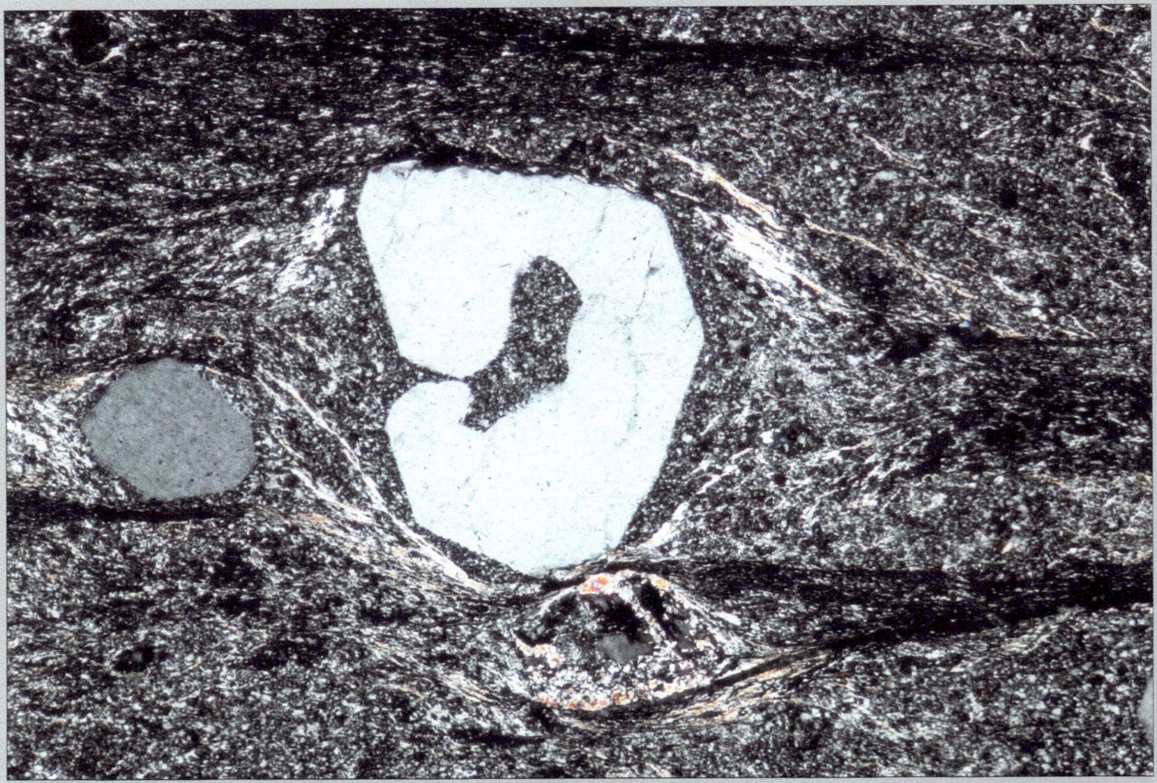

Chapter 12 | Exercises

12 Exercises

The purpose of this chapter is to give the reader the option to check his or her level of understanding of mylonites by trying to answer the questions given in the figure captions. Although we do not claim that our solutions given at the end of the chapter are the only possible ones, they will enable an assessment of the level of comprehension.

Fig. 12.1 Width of view 4 mm. PP.
Describe the structure, comment on the temperature during formation of the structure and determine sense of shear.

Fig. 12.2 Mylonite derived from granite. The central porphyroclast is feldspar and the high relief mineral in the upper central part is garnet. Pernambuco, NE Brazil. Width of view 3 mm. PPL.
Name the structure, commend on the temperature during mylonitisation and define the sense of shear.

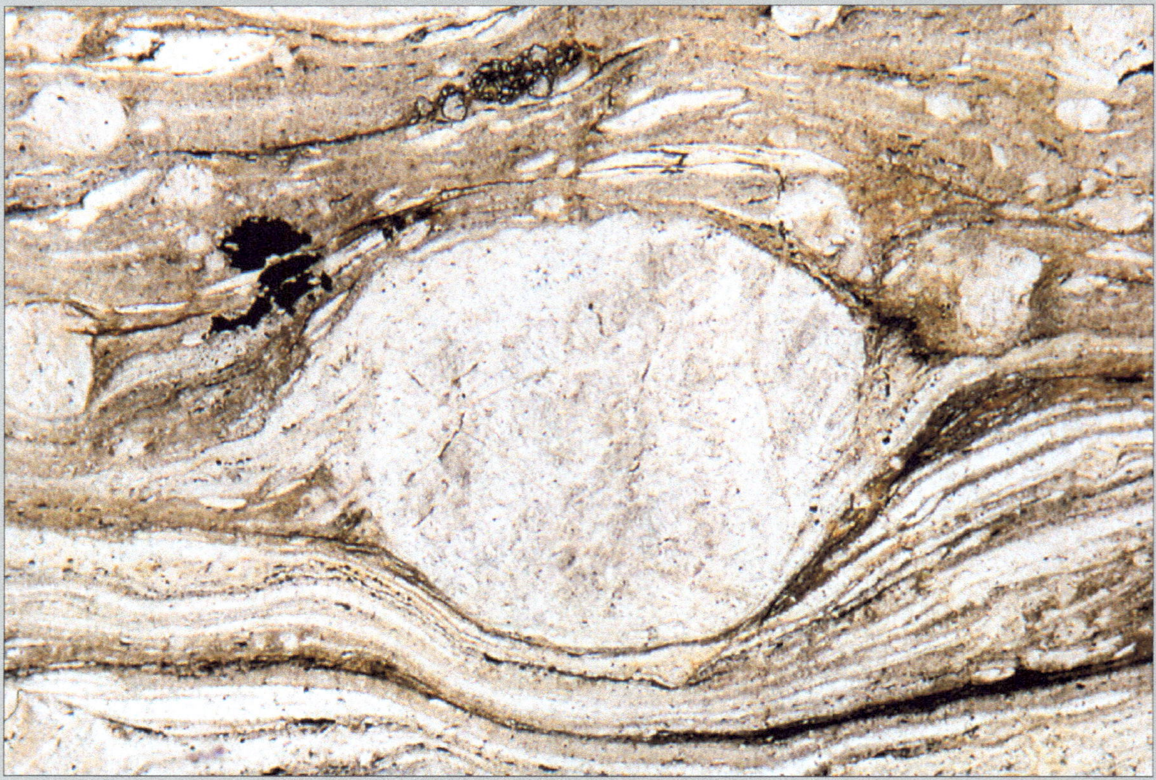

Fig. 12.3 Mylonite derived from quartz-mica schist. Heliodora, Southern Minas Gerais State, SE Brazil. Width of view 12 mm. CPL.
Describe the structure, define sense of shear and estimate the temperature during formation.

Fig. 12.4 Mylonite derived from a granitic rock. Mount West, Western Australia. Width of view 16 mm. PPL. Define the sense of shear and describe the structures used as criteria. Estimate the temperature during mylonitisation (see also Fig. 12.5).

Fig. 12.5 As Fig. 12.4 . Width of view 16 mm. CPL.

Fig. 12.6 Mylonite derived from a granitic rock. Mount West, Western Australia. Width of view 16 mm. PPL. Define the sense of shear and describe the structures used as criteria. Estimate the temperature during mylonitisation (see also Fig. 12.7).

Fig. 12.7 As Fig. 12.6 . Width of view 16 mm. CPL.

Fig. 12.8 Mylonite (lower part) and ultramylonite (upper part) derived from granite. Saint-Barthélemy Massif, French Pyrenees. Width of view 16 mm. PPL.
Determine the sense of shear, name the structures used as criteria and estimate the temperature during mylonitisation (see also Fig. 12.9).

Fig. 12.9 As Fig. 12.8. Width of view 16 mm. CPL.

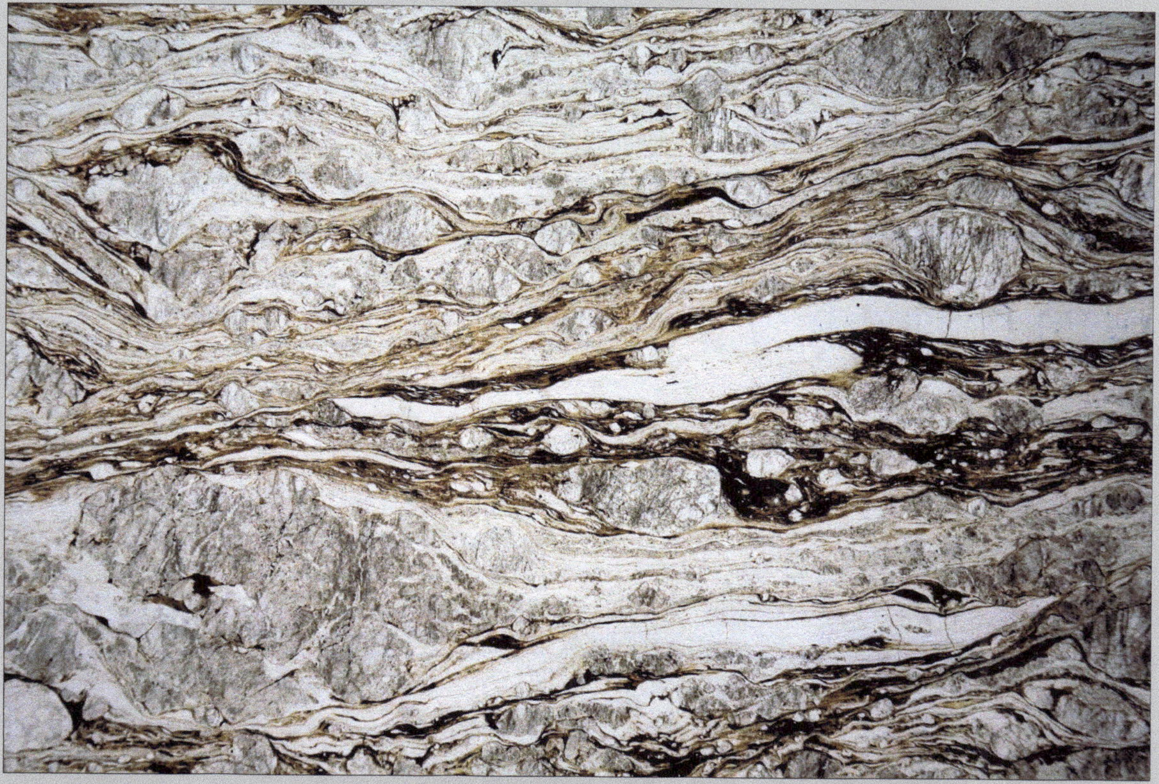

Fig. 12.10 Mylonite derived from granite. New Zealand. Width of view 30 mm. PPL.
Determine the sense of shear, name the structures used as criteria and estimate the temperature during mylonitisation (see also Fig. 12.11).

Fig. 12.11 As Fig. 12.10. Width of view 30 mm. CPL.

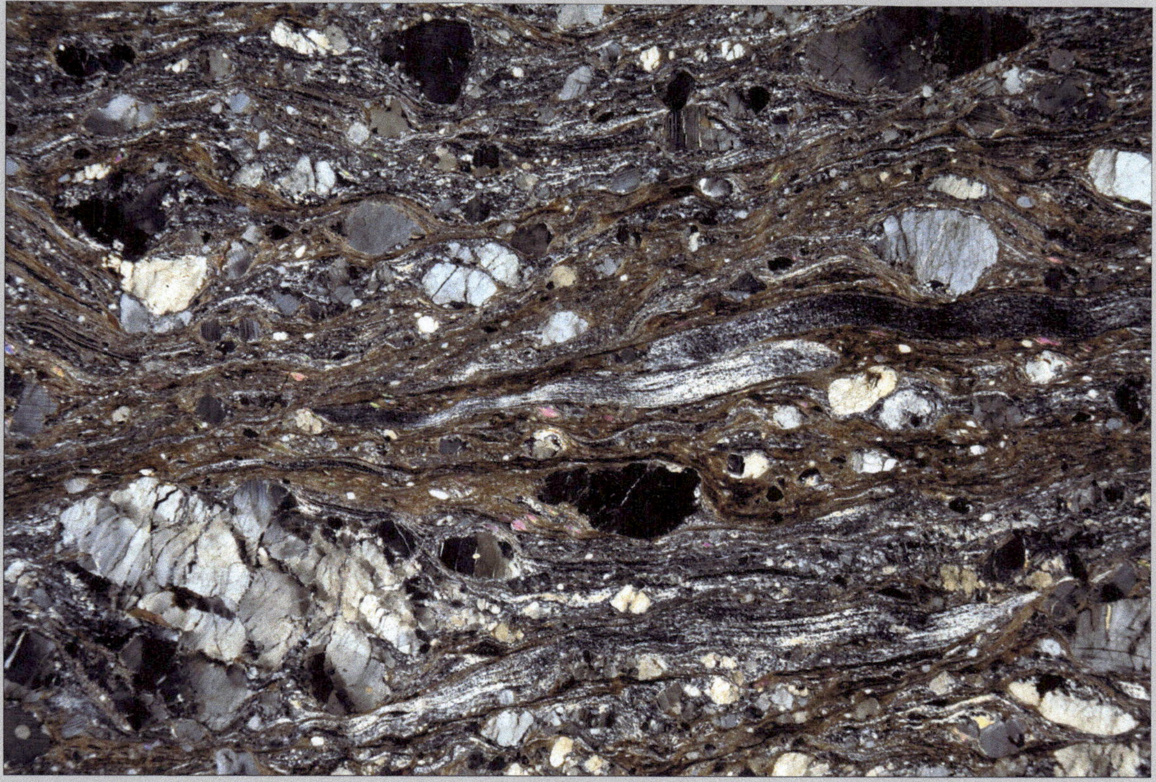

Fig. 12.12 Mylonite derived from a granitic rock. Rio Grande do Sul, southern Brazil. Width of view 16 mm. CPL. Determine the sense of shear, name the structures used as criteria and estimate the temperature during mylonitisation.

Fig. 12.13 Mylonite derived from a paragneiss. Pernambuco, NE Brazil. Width of view 22 mm. PPL. Determine the sense of shear, name the structures used as criteria and estimate the temperature during mylonitisation (see also Fig. 12.14).

Fig. 12.14 As Fig. 12.13. Width of view 22 mm. CPL.

Fig. 12.15 Mylonite derived from a garnet-rich paragneiss. Marsfjällen, Västerbotten, Sweden. Width of view 16 mm. PPL.
Determine the sense of shear, name the structures used as criteria and estimate the temperature during mylonitisation (see also Fig. 12.16).

Fig. 12.16 As Fig. 12.15. Width of view 16 mm. CPL.

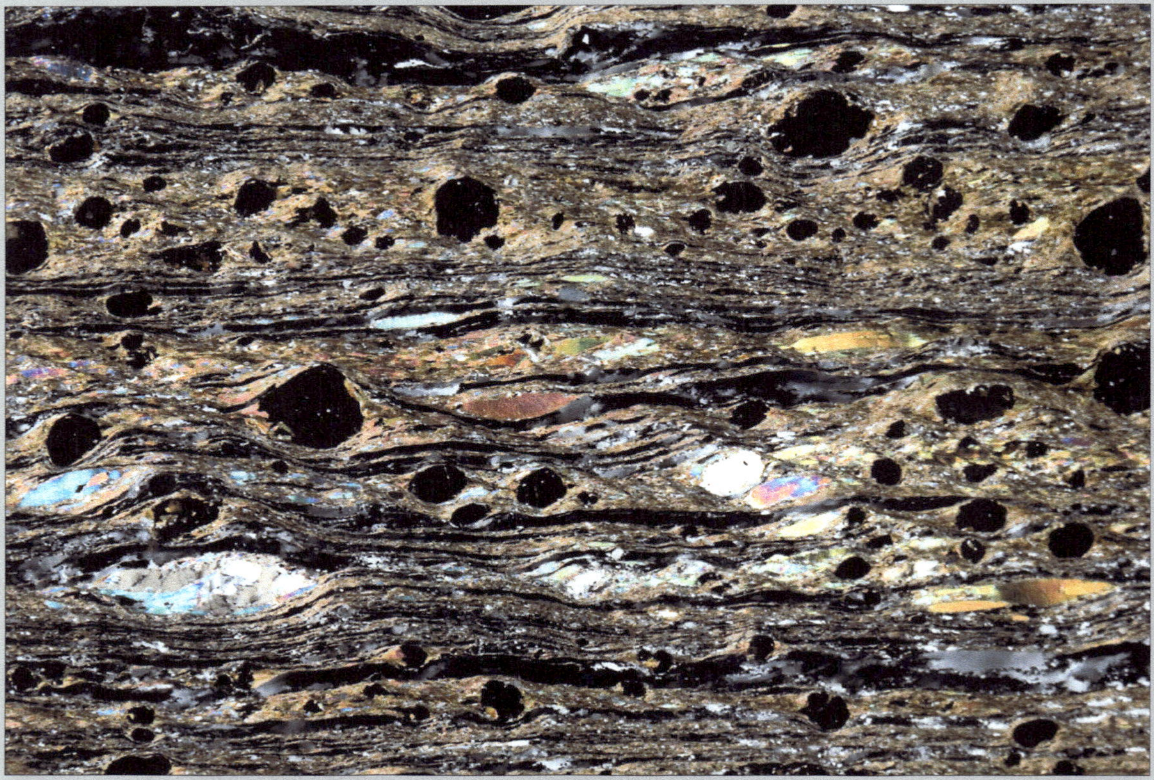

Fig. 12.17 Mylonite derived from a paragneiss. Pernambuco, NE Brazil. Width of view 8 mm. PPL. Determine the sense of shear, name the structures used as criteria and estimate the temperature during mylonitisation.

Fig. 12.18 Mylonite derived from a paragneiss. Pernambuco, NE Brazil. Width of view 4 mm. PPL. Determine the sense of shear, name the structures used as criteria and estimate the temperature during mylonitisation.

Fig. 12.19 Fish-shaped plagioclase grain embedded in a large quartz crystals (in extinction position). Varginha, southern Minas Gerais State, SE Brazil. Width of view 1.5 mm. CPL.
Determine the sense of shear, estimate the temperature during mylonitisation and comment on the structure.

Fig. 12.20 Three plagioclase fish and one hornblende fish (brownish, right hand side) embedded in large quartz crystals. Varginha, southern Minas Gerais State, SE Brazil. Width of view 3 mm. CPL.
Determine the sense of shear, estimate the temperature during mylonitisation and comment on the structure.

Fig. 12.21 Mylonite to ultramylonite derived from a granitic parent rock. Roraima, northern Brazil. Width of view 24 mm. PPL.
Determine the sense of shear, name the structures used as criteria and estimate the temperature during mylonitisation (see also Fig. 12.22).

Fig. 12.22 As Fig. 12.21. Width of view 24 mm. CPL.

Fig. 12.23 Mylonite derived from a garnet micaschist. Caxambu, southern Minas Gerais State, SE Brazil. Width of view 18 mm. PPL.
Determine the sense of shear, name the structures used as criteria and estimate the temperature during mylonitisation (see also Fig. 12.24).

Fig. 12.24 As Fig. 12.23. Width of view 18 mm. CPL.

Fig. 12.25 Mylonite derived from an orthogneiss. Pernambuco, NE Brazil. Width of view 18 mm. PPL.
Determine the sense of shear, name the structures used as criteria and estimate the temperature during mylonitisation (see also Fig. 12.26).

Fig. 12.26 As Fig. 12.25. Width of view 18 mm. CPL.

Fig. 12.27 Mylonite derived from a granitic parent rock. Rio Grande do Sul, southern Brazil. Width of view 12 mm. CPL.
Determine the sense of shear, name the structures used as criteria and estimate the temperature during mylonitisation.

Fig. 12.28 Quartz-rich part in a mylonitic gneiss. Leiden collection. Width of view 3 mm. CPL.
Determine the sense of shear, name the structure used as criterium and estimate the temperature during mylonitisation.

Fig. 12.29 Kyanite porphyroclast in a mylonitised biotite-rich mica schist. Caxambu, southern Minas Gerais State, SE Brazil. Width of view 3 mm. CPL.
Determine the sense of shear, name the structures used as criteria and estimate the temperature during mylonitisation.

Fig. 12.30 Mylonitised quartz-mica schist. Caxambu, southern Minas Gerais State, SE Brazil. Width of view 3 mm. CPL.
Determine the sense of shear, name the structures used as criteria and estimate the temperature during mylonitisation.

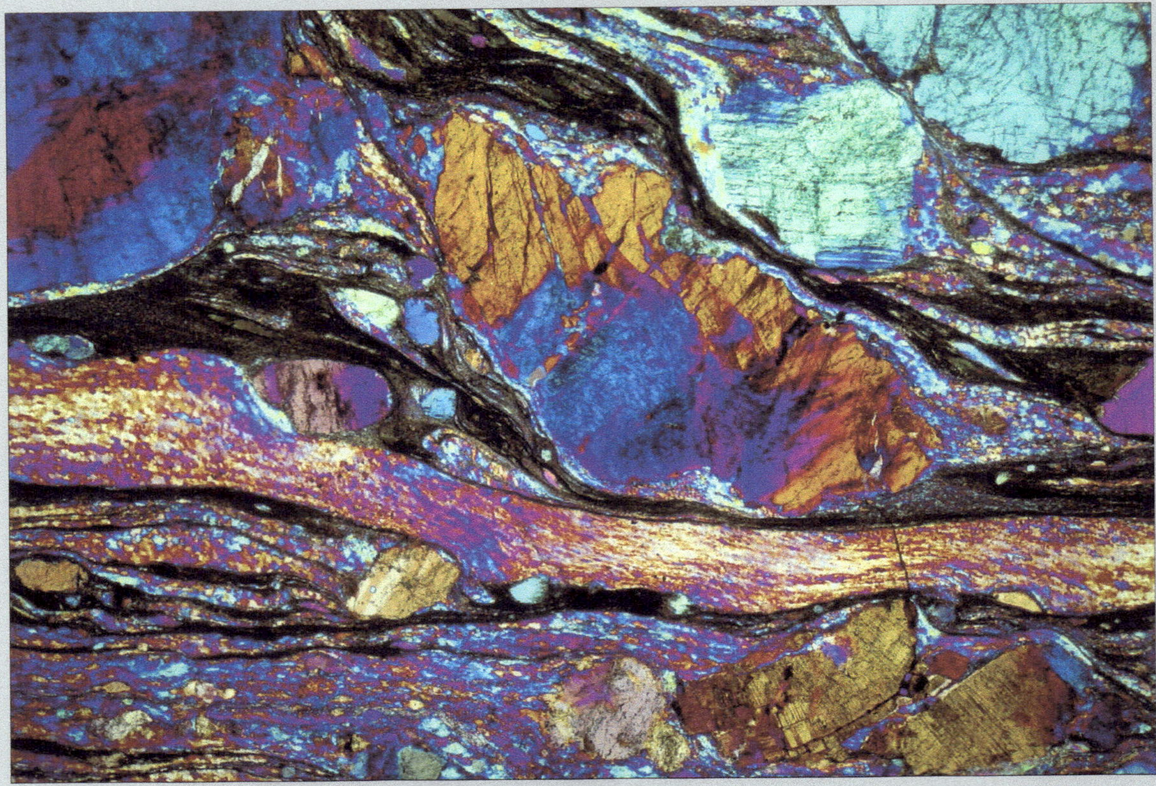

Fig. 12.31 K-feldspar porphyroclast in a mylonite derived from a granitic rock. South Island, New Zealand. Width of view 4 mm. CPL with gypsum plate.
Determine the sense of shear, name the structures used as criteria and estimate the temperature during mylonitisation.

Fig. 12.32 Mylonite derived from feldspar-bearing mica schist. Caxambu, southern Minas Gerais State, SE Brazil. Width of view 5 mm. CPL.
Determine the sense of shear, name the structures used as criteria and estimate the temperature during mylonitisation.

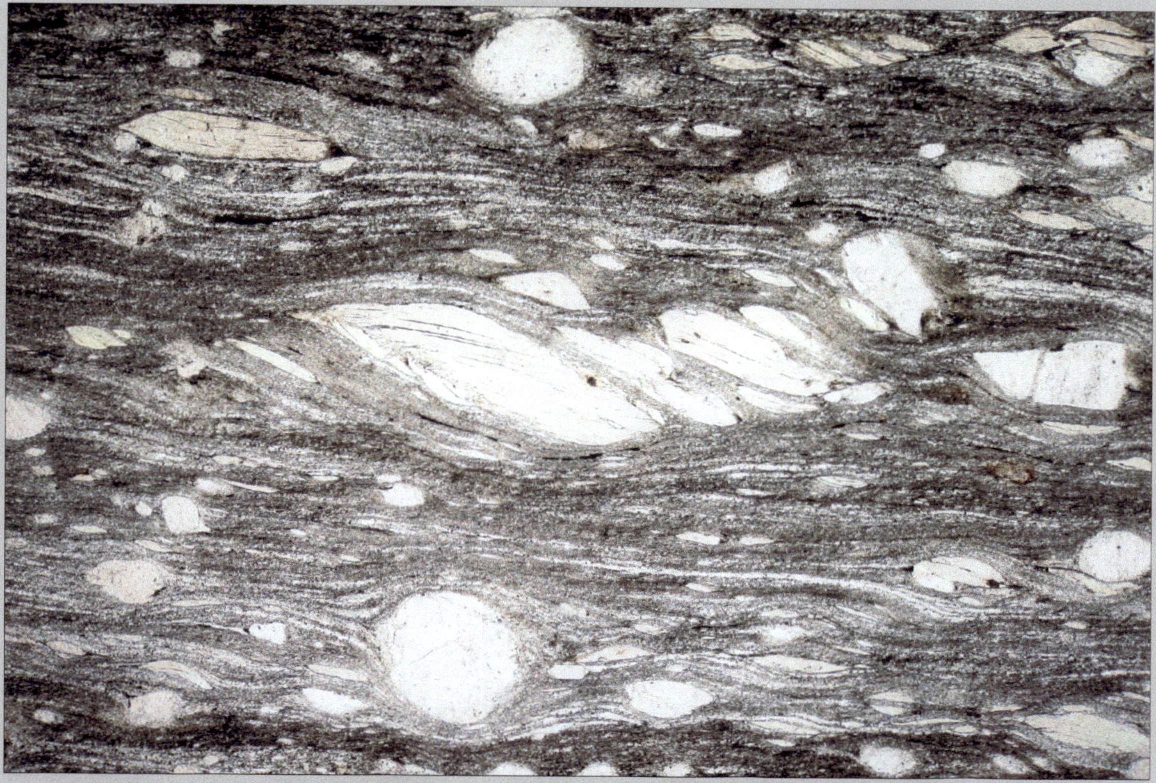

Fig. 12.33 Mylonite derived from feldspar-bearing mica schist. Caxambu, southern Minas Gerais State, SE Brazil. Width of view 5 mm. CPL.

Determine the sense of shear, name the structures used as criteria and estimate the temperature during mylonitisation (see also Fig. 12.34).

Fig. 12.34 As Fig. 12.33. The field covered by this photomicrograph is slightly different from Fig. 12.33, but the major structure in the center can be recognized on both. Width of view 5 mm. CPL.

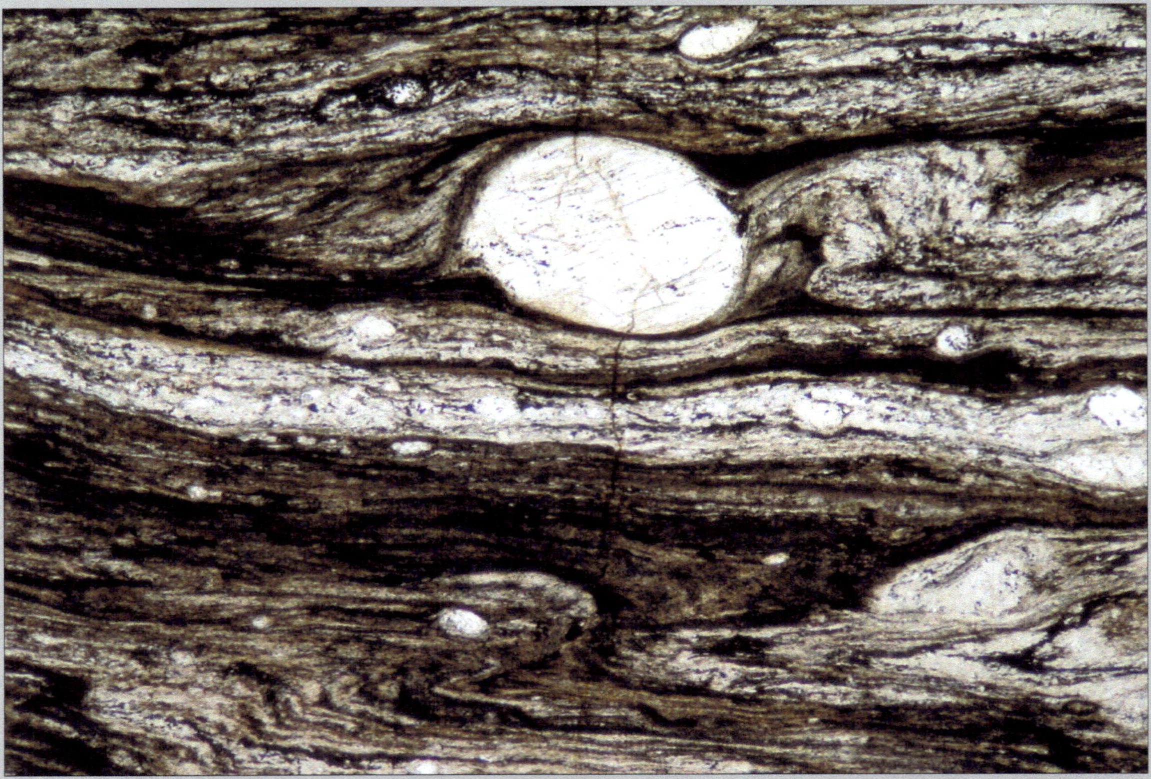

Fig. 12.35 Mylonite derived from granitic rock, with feldspar porphyroclast. Roraima, northern Brazil. Width of view 2 mm. PPL.
Determine the sense of shear and name the structures used as criteria.

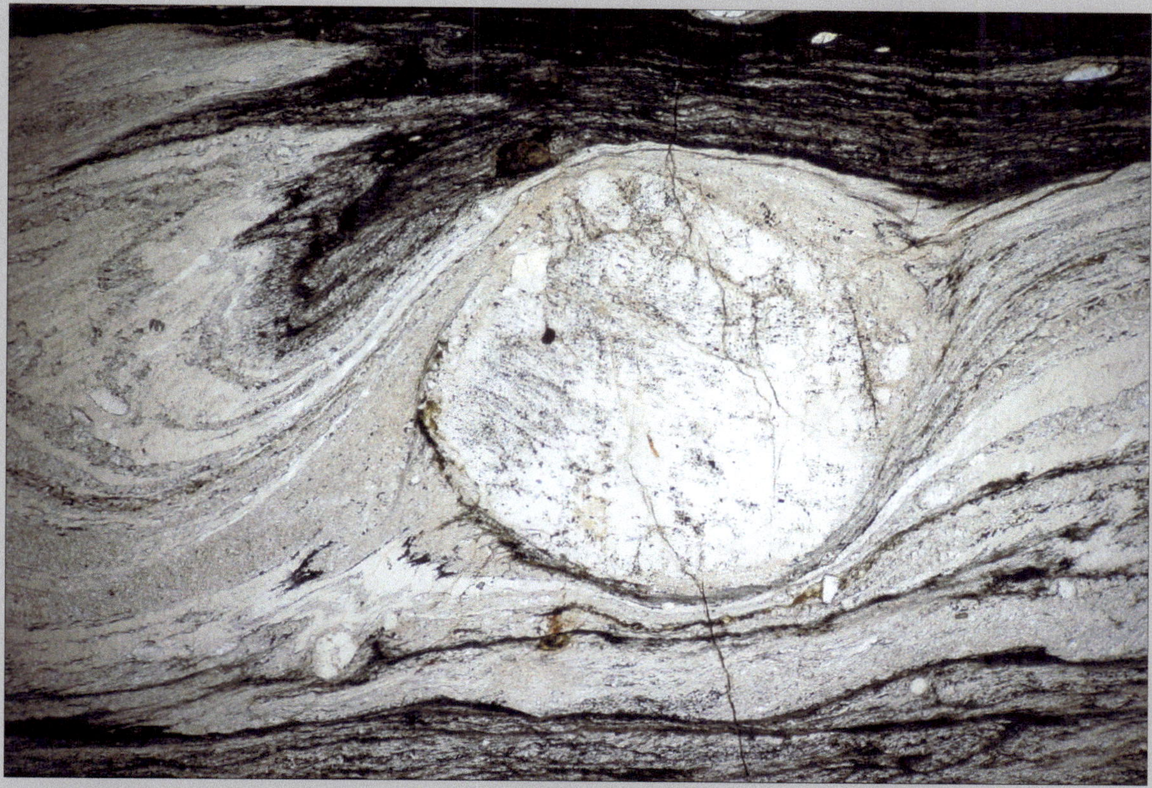

Fig. 12.36 Mylonite derived from granitic rock with large porphyroclast of K-feldspar. Roraima, northern Brazil. Width of view 5 mm. PPL.

Determine the sense of shear, name the structures used as criteria and estimate the temperature during mylonitisation (see also Fig. 12.37).

Fig. 12.37 As Fig. 12.36. Width of view 5 mm. CPL.

Fig. 12.38 Mylonite derived from a paragneiss with muscovite and K-feldspar porphyroclasts. Pernambuco, NE Brazil. Width of view 3.2 mm. CPL.
Determine the sense of shear, name the structures used as criteria and estimate the temperature during mylonitisation.

Fig. 12.39 Mylonite derived from a paragneiss with muscovite and K-feldspar porphyroclasts. Pernambuco, NE Brazil. Width of view 2.5 mm. CPL.
Determine the sense of shear, name the structures used as criteria and estimate the temperature during mylonitisation.

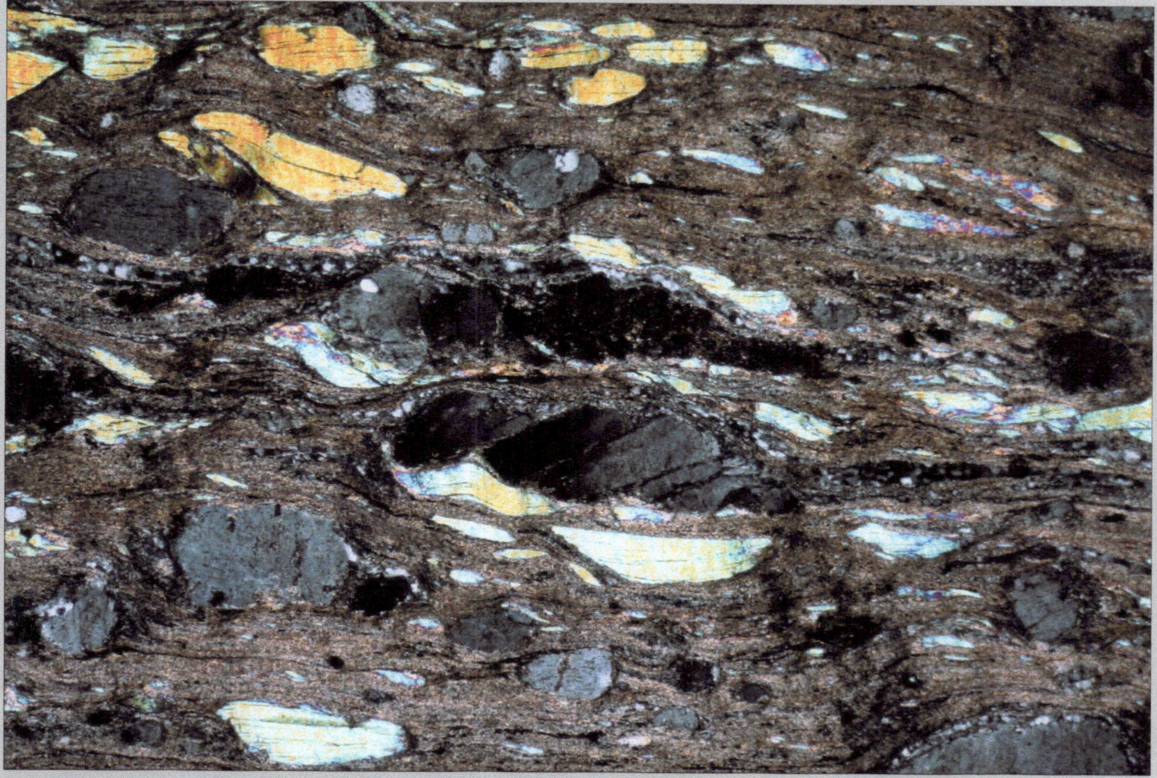

Fig. 12.40 Mylonite derived from a paragneiss with muscovite and K-feldspar porphyroclasts. Pernambuco, NE Brazil. Width of view 2 mm. CPL.
Determine the sense of shear, name the structures used as criteria and estimate the temperature during mylonitisation.

Fig. 12.41 Mylonite derived from granite with large plagioclase porphyroclast. Saint Barthélemy Massif, French Pyrenees. Width of view 5 mm. CPL.
Determine the sense of shear, name the structures used as criteria and estimate the temperature during mylonitisation.

Fig. 12.42 Ultramylonite with fine-grained quartz and biotite. Saint-Barthélemy Massif, French Pyrenees. Width of view 3 mm. CPL.
Determine the sense of shear, name the structures used as criteria and estimate the temperature during mylonitisation.

Fig. 12.43 Mylonite with hornblende porphyroclast in a matrix composed of fine-grained feldspar and coarser-grained quartz. Miguel Pereira, Rio de Janeiro State, SE Brazil. Width of view 14 mm. PPL.
Determine the sense of shear, name the structures used as criteria and estimate the temperature during mylonitisation (see also Fig. 12.44).

Fig. 12.44 As Fig. 12.43. Width of view 14 mm. CPL.

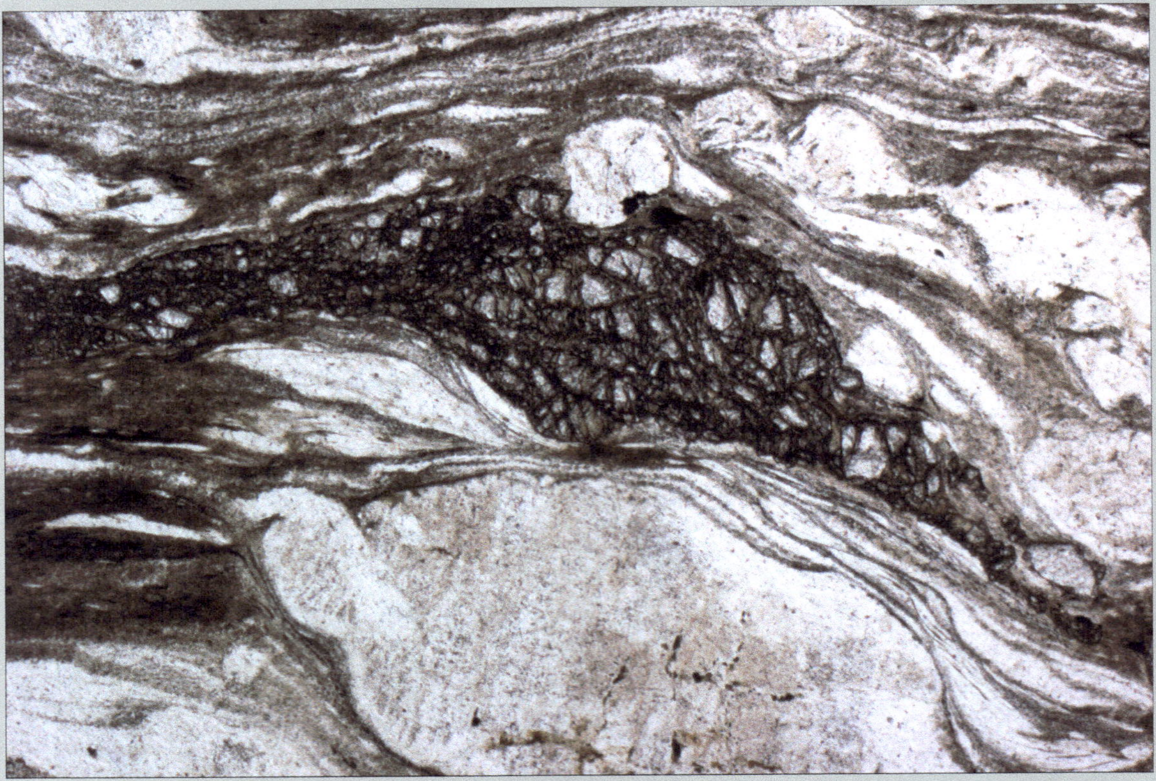

Fig. 12.45 Mylonite derived from paragneiss with feldspar, muscovite and garnet porphyroclasts. Pernambuco, NE Brazil. Width of view 4 mm. PPL.
Determine the sense of shear, name the structures used as criteria and estimate the temperature during mylonitisation.

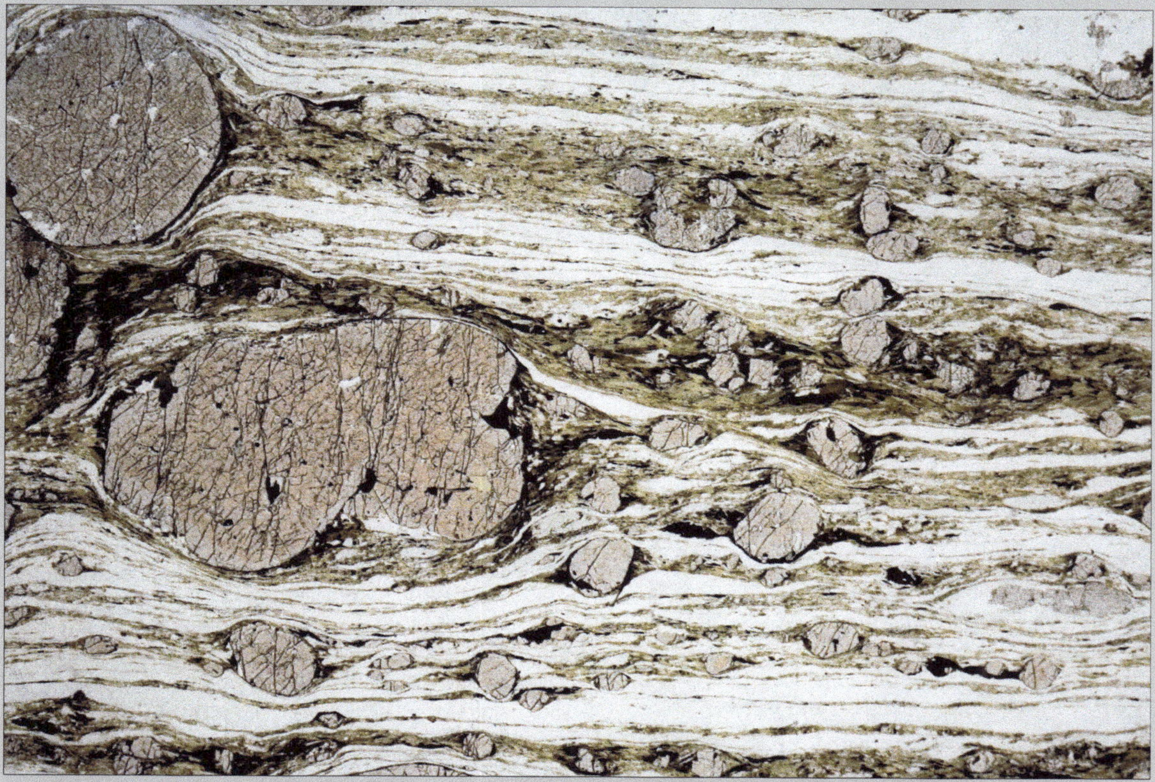

Fig. 12.46 Mylonite derived from garnet mica schist. Marsfjällen, Västerbotten, Sweden. Width of view 16 mm. PPL.

Determine the sense of shear, name the structures used as criteria and estimate the temperature during mylonitisation (see also Fig. 12.47).

Fig. 12.47 As Fig. 12.46. Width of view 16 mm. CPL.

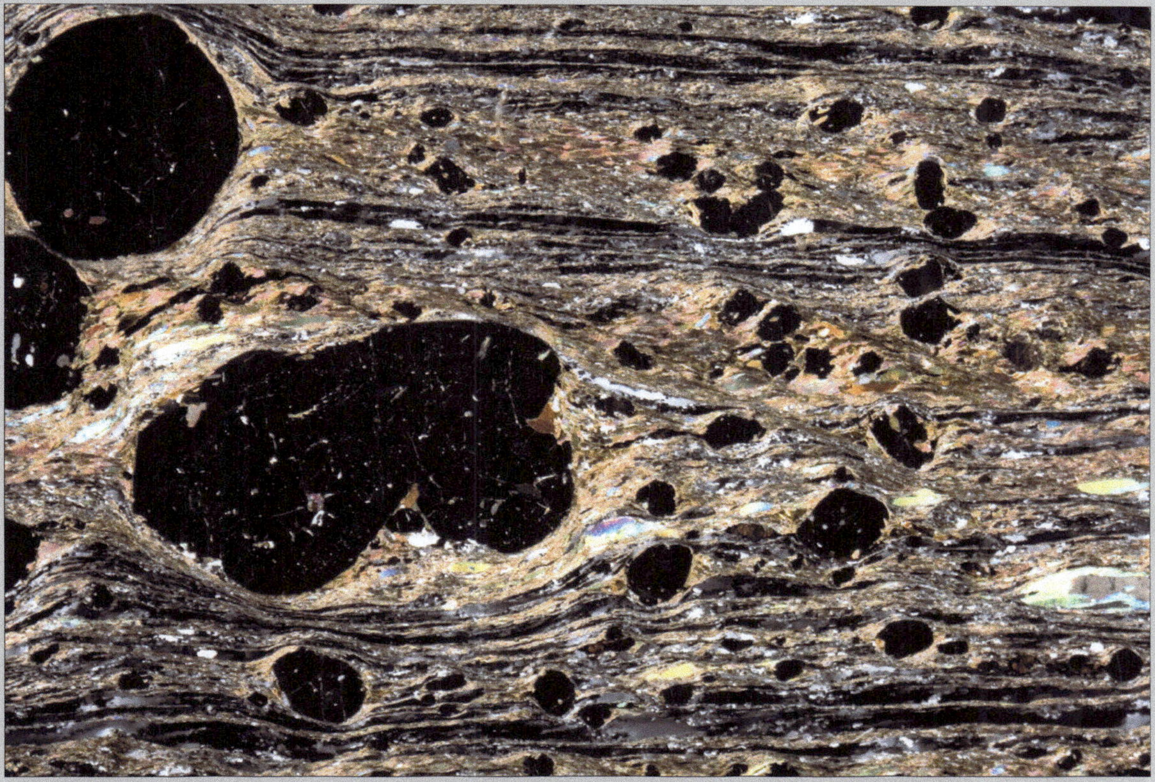

Fig. 12.48 Mylonite derived from paragneiss with muscovite and feldspar porphyroclasts. Pernambuco, NE Brazil. Width of view 16 mm. PPL.
Determine the sense of shear, name the structures used as criteria and estimate the temperature during mylonitisation (see also Fig. 12.49).

Fig. 12.49 As Fig. 12.48. Width of view 16 mm. CPL.

Fig. 12.50 Mylonite derived from paragneiss with porphyroclasts of muscovite and feldspar embedded in a matrix composed of fine-grained quartz and mica. Pernambuco, NE Brazil. Width of view 4 mm. CPL. Saint Barthélemy Massif, French Pyrenees. Width of view 5 mm. CPL.
Determine the sense of shear, name the structures used as criteria and estimate the temperature during mylonitisation.

Fig. 12.51 Mylonite derived from a meta-rhyolite with porphyroclasts of quartz embedded in a matrix of fine-grained quartz and mica. Northeast Minas Gerais State, SE Brazil. Width of view 18 mm. CPL.
Determine the sense of shear, name the structures used as criteria and estimate the temperature during mylonitisation.

Fig. 12.52 Mylonitic schist with foliation fish composed of muscovite in a matrix of fine-grained quartz and mica. Santana do Garambeu, southern Minas Gerais State, SE Brazil. Width of view 16 mm. CPL.
Determine the sense of shear, name the structures used as criteria and estimate the temperature during mylonitisation.

Interpretations

Fig. 12.1 The dark rock occupying most of the photomicrograph is a pseudotachylyte as is evident from the injection veins along the upper contact. Within the pseudotachylyte a foliation can be recognized defined by elongated inclusions due to ductile deformation superimposed on the pseudotachylyte formation. The angle of this foliation with the more inclined pseudotachylyte contacts suggests dextral sense of shear for the ductile deformation. The temperature must have been low during pseudotachylyte formation but must have been higher during the ductile deformation.

Fig. 12.2 The structure around the central porphyroclast can be defined as a delta structure, that, together with well defined stair stepping, points to dextral sense of shear. The metamorphic conditions during mylonitisation were low grade because of the broken garnet in the central upper part of the photomicrograph and because of the very fine-grained matrix.

Fig. 12.3 The structures indicative for sense of shear in this thin section are mica fish (e.g. along the lower limit of the photomicrograph) and C′type shear bands (from upper right to lower left). Both indicate sinistral sense of shear. The undulose extinction of muscovite points to relatively low-grade metamorphic conditions during and after mylonitisation.

Figs. 12.4 and 12.5 Shear sense indicators are asymmetric folds (lower left and just below the center) verging top to the left; C/S fabric (C almost horizontal and S inclined to the right) and a C′shear band (top right to bottom left, in the central part), all indicating sinistral sense of shear. The metamorphic conditions during mylonitisation were low- to medium-grade from the partial recrystallisation of feldspar around porphyroclasts, forming core-and-mantle structures (e.g. in the center and top right).

Figs. 12.6 and 12.7 A moderately developed C/S fabric (better visible with CPL, C planes subhorizontal, marked as dark zones, ans S planes inclined to the right) and a porphyroclast at right with a core-mantle structure and slight stair stepping, mark the sense of shear as sinistral. The partial recrystallisation of the feldspar porphyroclast at right and the elongated deformed quartz grains indicate low- to medium-grade metamorphic conditions.

Figs. 12.8 and 12.9 In the upper ultramylonitic part a delta clast can be observed at the right hand side in the PPL photo, indicating dextral sense of shear. Along the lower limit of the photomicrograph the mylonitic foliation is slightly inclined to the left, representing S planes. These also indicate dextral sense of shear with respect to the subhorizontal ultramylonite foliation (closer to C planes). The metamorphic conditions must have been low grade as can be inferred from the strongly elongated quartz grains (close to the bottom, CPL) and the apparent lack of recrystallisation around feldspar porphyroclasts (right hand side).

Figs 12.10 and 12.11 Several asymmetric folds (a small one just above the center and an isoclinal one just right of the center) indicate sinistral sense of shear, reinforced by an oblique C′type shear band that runs from top right to center left. Several porphyroclast systems show slight stair stepping to the left. The temperature was relatively low, judging from the small grain size of the partially recrystallised quartz (e.g. folded vein right from center and lower right). The feldspars are partially broken and do not show significant recrystallisation at this magnification.

Fig. 12.12 The lozenge shaped perthitic K-feldspar porphyroclast in the upper left corner indicates sinistral sense of shear according to the criteria of mineral fish and stair stepping. Around this porphyroclast a C/S fabric can be recognized with subhorizontal C-planes and S-planes inclined to the right, also indicating sinistral shear sense. The metamorphic conditions were probably low- to medium-grade during mylonitisation judging from the partial recrystallisation in the rim of the K-feldspar porphyroclast.

Figs. 12.13 and 12.14 A large number of mica fish visible with CPL (Fig. 12.14), many of which are slightly inclined to the right, indicate sinistral sense of shear (especially in the upper left corner). In Fig. 12.13 several porphyroclast systems with stair stepping to the left (e.g. in the lower left corner) and a small asymmetric fold (lower right) confirm this sense of shear. The metamorphic conditions during mylonitisation are difficult to estimate at this magnification but the lack of significant recrystallisation in the feldspar porphyroclasts is consistent with low grade.

Figs. 12.15 and 12.16 Mica and kyanite fish slightly inclined to the left (e.g. at left just below the dividing line of upper and lower parts) and C′type shear bands inclined to the right (e.g. next to major garnet at left) indicate dextral sense of shear. Quartz ribbons with strong undulose extinction and limited recrystallisation (lower right) are characteristic for low-grade metamorphic conditions during mylonitisation.

Fig. 12.17 Several mica fish at the lower left hand side, slightly inclined to the right, indicate a sinistral sense of shear, confirmed by porphyroclasts with stair stepping to the left (e.g. upper left). No metamorphic grade can be deduced from this photo but low- to medium-grade is probable because of the small grain size of the matrix.

Fig. 12.18 Although the longest dimension of the central porphyroclast is inclined to the right, stair stepping of the deviating mylonitic foliation indicates clearly sinistral sense of shear, confirmed by several mica fish (below center and upper right) inclined to the right. Just above the central porphyroclast is a string of garnet fragments, probably derived from a cataclased larger grain, indicating low-grade metamorphic conditions during mylonitisation.

Fig. 12.19 The perfect plagioclase fish, embedded in coarse grained quartz, indicates sinistral sense of shear. After mylonitisation the temperature was sufficiently high to permit grain growth in quartz by high-temperature grain boundary migration. The fish remains as a relict of mylonitic structure in a quartzitic part of a high-grade gneiss (not visible in the photomicrograph).

Fig. 12.20 The two large plagioclase fish both indicate sinistral sense of shear; note that the reference plane is slightly inclined to the left. The hornblende fish at the right hand side is truncated by the side of the photograph, but its shape is consistent with sinistral shear as well. The large quartz grains show lobate contacts, typical for growth by high-grade grain boundary migration. The black grain at upper left is a garnet fish and the dark spots "raining" out of it are fluid inclusions.

Figs. 12.21 and 12.22 Dextral sense of shear is indicated by tight asymmetric folds, verging to the right, by accentuated stair stepping across most porphyroclasts and by a C/S fabric, with subhorizontal C planes and S planes inclined to the left. The porphyroclasts show very little recrystallisation along the rims, indicating low-grade metamorphic conditions.

Figs. 12.23 and 12.24 The inclined muscovite fish in the lower left quadrant indicates sinistral sense of shear. The recrystallised quartz vein (just above the center) is indicative for medium-grade metamorphic conditions.

Figs. 12.25 and 12.26 Sinistral shear sense indicators include mica fish (Fig. 12.26), a delta-shaped folded quartz vein (right of center), a sigma-type porphyroclast system at upper right and several porphyroclasts with subtle stair stepping to the left. The metamorphic conditions are difficult to judge from these images; they were probably low- to medium-grade because of the fine-grained matrix.

Fig. 12.27 A typical C/S fabric, with C planes inclined to the left and S planes inclined to the right, indicates sinistral sense of shear, confirmed by several mica fish (upper right) and stair stepping across porphyroclasts (below and left of center). Metamorphic conditions were probably medium grade judging by the partial recrystallisation of porphyroclasts.

Fig. 12.28 The oblique foliation in quartz, steeply inclined to the right, indicates sinistral sense of shear. The deformed quartz with undulose extinction and lobate contacts points to low-grade metamorphic conditions during mylonitisation.

Fig. 12.29 The kyanite fish in the center (orange to purple) and several mica fish left of it reveal sinistral sense of shear. The quartz in the upper part of the photo is well recrystallised to a granoblastic polygonal texture, typical for medium-grade metamorphic conditions.

Fig. 12.30 The mica fish indicate sinistral sense of shear. The granoblastic polygonal quartz texture, resulting from recrystallisation, is characteristic for medium-grade metamorphic conditions.

Fig. 12.31 Sinistral sense of shear is indicated by the oblique foliation in the quartz band, just below the center. The fragmented feldspar porphyroclast at lower right (yellow) shows a synthetic microfault with apparent sinistral displacement, reforcing the interpretation of sinistral sense of shear. The strongly deformed quartz with only incipient recrystallisation and the fragmented feldspar at lower right indicate low-grade metamorphic conditions.

Fig. 12.32 The mica fish are subhorizontal to slightly inclined to the right, indicating sinistral sense of shear. A feldspar porphyroclast at lower right also indicates sinistral shear sense by stair stepping. The quartz in the vein close to the bottom is well recrystallised to a polygonal granoblastic aggregate and the feldspar porphyroclast at lower right shows minor recrystallisation along its upper part, hence the metamorphic conditions can be estimated as medium grade.

Figs. 12.33 and 12.34 The oblique mica fish throughout the image indicate sinistral sense of shear. The well recrystallised quartz vein in the upper part of Fig. 12.34 points to medium-temperature conditions during mylonitisation.

Fig. 12.35 The main kinematic indicators in this image are asymmetric folds verging to the right (lower left and upper right). The feldspar porphyroclast just above the center shows stair stepping of a dark band in the matrix. Both indicators point to dextral sense of shear.

Figs. 12.36 and 12.37 The kinematic indicators that can be recognized in this image are oblique foliation (left hand part) asymmetric folds verging to the right (upper left), quarter folds and a delta structure around the central porphyroclast. All indicate dextral sense of shear. The elongate deformed quartz (left of center) is diagnostic for low-grade metamorphic conditions.

Fig. 12.38 The central K-feldspar porphyroclast shows stair stepping to the left and mica fish at upper right also indicate sinistral sense of shear. At lower left of the porphyroclast is a fragmented porphyroclast with domino-type antithetic dextral offset, consistent with sinistral displacement in the rock. This fragmented porphyroclast and the small, partially recrystallised quartz left and right of the central porphyroclast indicate low temperature.

Fig. 12.39 Sinistral sense of shear is indicated by shear-band-type offset in the central fragmented porphyroclast and also by two mica fish at lower left and by the stair stepping across the central porphyroclast. The presence of the fragmented porphyroclast points to low-grade metamorphic conditions.

Fig. 12.40 Several mica fish, especially at upper left, indicate sinistral sense of shear. Just below the center is a shear band type fragmented porphyroclast with sinistral internal displacement, synthetic to the main movement in the mylonite. The presence of this porphyroclast points to low-grade metamorphic conditions.

Fig. 12.41 Oblique foliation in the quartz-rich part to the right, a kind of duplex structure in this same part and minor sinistral dislocation along subhorizontal faults in the large plagioclase porphyroclast at lower left, all indicate sinistral sense of shear. Elongate deformed quartz with undulose extinction (upper right) and the fragmented plagioclase point to low-grade metamorphic conditions. Note a minor fault at upper left with sinistral offset that seems post mylonitisation.

Fig. 12.42 The oblique foliation in the quartz-rich bands, inclined to the left, and the asymmetric folds, verging to the right, both indicate dextral sense of shear. The quartz is partially recrystallised to small new grains, characteristic for low-grade metamorphic conditions.

Figs. 12.43 and 12.44 The hornblende porphyroclast shows asymmetric strain shadows with well defined stair stepping to the left, indicating sinistral shear sense. The recrystallised feldspar in the matrix and the large quartz grains, due to growth by high-temperature grain boundary migration, point to high-grade metamorphic conditions during and after the mylonitisation.

Fig. 12.45 Several mica fish inclined to the right and stair stepping around a major porphyroclast in the lower part of the photomicrograph indicate sinistral sense of shear. The fragmented garnet porphyroclast in the center also shows stair stepping to the left confirming the sense of shear. Metamorphic conditions were probably of low grade, judging from the fragmented garnet crystal.

Figs. 12.46 and 12.47 Shear sense indicators are difficult to define in this set of photomicrographs. However, at lower right is a mica fish and another one with kyanite inside, indicating dextral sense of shear. Two vague C′type shear bands, one in the center and one to the right also point to dextral sense of shear. The metamorphic grade was low, as can be deduced from the strongly elongated quartz grains with undulose extinction.

Figs. 12.48 and 12.49 Sinistral shear sense is indicated by a sigma-type porphyroclast right of the center and stair stepping to the left across a porphyroclast system close to the upper limit of the photomicrograph. Several mica fish and a minor asymmetric fold upper right of the center complete the sinistral shear sense indicators. Recrystallisation of quartz to minor equidimensional grains reveals low to medium temperature during mylonitisation.

Fig. 12.50 The fragmented porphyroclast in the center can be interpreted as domino type, indicating sinistral sense of shear, or as shear band type, indicating dextral shear sense. However, mica fish in the matrix around the porphyroclast (upper right) are consistent with sinistral shear sense, hence the first interpretation is preferred. Small, partially recrystallised quartz at upper right of the fragmented porphyroclast points to low-grade metamorphic conditions.

Fig. 12.51 The oblique position of the quartz fish, inclined to the right, and the wavy pattern of the mylonitic foliation, caused by C′-type shear bands inclined to the left, both point to sinistral sense of shear. Low-grade metamorphic conditions are indicated by the deformed quartz crystals with clear undulose extinction.

Fig. 12.52 The foliation fish of muscovite show dextral sense of shear and the recrystallised quartz grains in the matrix reveal medium-temperature conditions during mylonitisation.

References · Bibliography

References

Blenkinsop TG (2000) Deformation Microstructures and Mechanisms in Minerals and Rocks. Kluwer Academic Publishers, Dordrecht, 150 pp

Lapworth C (1885) The highland controversy in British geology: its causes, course and consequences. Nature 32:558–559

Mawer CK (1986) What is a mylonite? Geoscience Canada 13:33–34

Newman J, Lamb WM, Drury MR, Vissers RLM (1999) Deformation processes in a peridotite shear zone: reaction-softening by an H_2O-deficient, continuous net transfer reaction. Tectonophysics 303:193–222

Passchier CW, Trouw RAJ (2005) Microtectonics. 2nd edition, Springer Verlag, Heidelberg, 366 pp

Ramsay JG (1980a) Shear zone geometry: a review. J Struct Geol 2:83-101

Schmid SM, Handy MR (1991) Towards a genetic classification of fault rocks: geological usage and tectonophysical implications. In: Müller DW, McKenzie JA, Weissert H (eds) Controversies in Modern Geology, Evolution of Geological Theories in Sedimentology, Earth History and Tectonics. Academic Press, London, pp 339–361

Scholz C H (1990) The mechanics of earthquakes and faulting. Cambridge University Press, New York, 439 pp

Shelley D (1993) Igneous and metamorphic rocks under the microscope. Chapman and Hall, London

Sibson RH (1977b) Fault rocks and fault mechanisms. J Geol Soc Lond 133:191–213

Snoke A, Tullis J (1998) An overview of fault rocks. In: Snoke A, Tullis J, Todd VR (eds) Fault related rocks – a photographic atlas. Princeton University Press, New Jersey. pp 3-18

Snoke A, Tullis J, Todd VR (1998) Fault related rocks – a photographic atlas. Princeton University Press, New Jersey. 617 pp

Spray JG (1992) A physical basis for the frictional melting of some rock-forming minerals. Tectonophysics 204:205–221

Ten Grotenhuis SM, Trouw RAJ, Passchier CW (2003) Evolution of mica fish in mylonitic rocks. Tectonophysics 372:1–21

Vernon RH (2004) A practical guide to rock microstructures. Cambridge University Press, 594 pp

Wise DU, Dunn DE, Engelder JT, Geiser PA, Hatcher RD, Kish SA, Odom AL, Schamel S (1984) Fault-related rocks: Suggestions for terminology. Geology 12:391–394

Bibliography

Allen AR (1979) Mechanism of frictional fusion in fault zones. J Struct Geol 1:231–243

Allison I, Barnett RL, Kerrich R (1979) Superplastic flow and changes in crystal chemistry of feldspars. Tectonophysics 53:41–46

Alsop GI, Carreras J (2007) The structural evolution of sheath folds: A case study from Cap de Creus. J Struct Geol 29:1915–1930

Alsop GI, Holdsworth RE (2004) The geometry and topology of natural sheath folds: a new tool for structural analysis. J Struct Geol 26:1561–1589

Alsop GI, Holdsworth RE (2006) Sheath folds as discriminators of bulk strain type. J Struct Geol 28:1588–1606

Alsop GI, Holdsworth RE, McCaffrey KJW (2007) Scale invariant sheath folds in salt, sediments and shear zones. J Struct Geol 29:1585–1604

Antonellini MA, Aydin A, Pollard DA (1994) Microstructure of deformation bands in porous sandstones at Arches National Park, Utah. J Struct Geol 16:941–959

Arancibia G, Morata D (2005) Compositional variations of syntectonic white-mica in low-grade ignimbritic mylonite. J Struct Geol 27:745–767

Arbaret L, Diot H, Bouchez JL (1996) Shape fabrics of particles in low concentration suspensions; 2D analogue experiments and application to tiling in magma. J Struct Geol 18:941–950

Archanjo CJ, Hollanda MHBM, Rodrigues SWO, Neves BBB, Armstrong R (2008) Fabrics of pre- and syntectonic granite plutons and chronology of shear zones in the Eastern Borborema Province, NE Brazil. J Struct Geol 30:310–326

Arslan A, Passchier CW, Koehn D (2008) Foliation boudinage. J Struct Geol 30:291–309

Ashley Griffith W, Di Toro G, Pennacchioni G, Pollard DD (2008) Thin pseudotachylytes in faults of the Mt. Abbot quadrangle, Sierra Nevada: Physical constraints for small seismic slip events. J Struct Geol 30:1086–1094

Austin NJ, Evans B (2007) Paleowattmeters: A scaling relation for dynamically recrystallized grain size. Geology 35:343–346

Austrheim H, Boundy TM (1994) Pseudotachylytes generated during seismic faulting and eclogitization of the deep crust. Science 265:82–83

Aydin A (1978) Small faults formed as deformation bands in sandstone. Pure Appl Geophys 116:913–930

Aydin A (2000) Fractures, faults, and hydrocarbon entrapment, migration and flow. Mar Petrol Geol 17:797–814

Aydin A, Johnson AM (1978) Development of fault zones as zones of deformation bands and as slip surfaces in sandstone. Pure Appl Geophys 116:931–942

Aydin A, Johnson AM (1983) Analysis of Faulting in porous sandstones. J Struct Geol 5:19–31

Azor A, Fernando Simancas J, Exposito I, Lodeiro FG, Martinez Poyatos DJ (1997) Deformation of garnets in a low-grade shear zone. J Struct Geol 19:1137–1148

Babaie HA, La Tour TE (1994) Semibrittle and cataclastic deformation of hornblende-quartz rocks in a ductile shear zone. Tectonophysics 229, pp 19–30

Babaie HA, La Tour TE (1998) Semi-brittle deformation and cataclastic flow of hornblende-quartz rock in a ductile shear zone. In: Snoke A, Tullis J, Todd VR (eds) Fault related rocks – a photographic atlas. Princeton University Press, New Jersey:180–183

Baird GB, Hudleston PJ (2007) Modeling the influence of tectonic extrusion and volume loss on the geometry, displacement, vorticity, and strain compatibility of ductile shear zones. J Struct Geol 29:1665–1678

Bak J, Korstgard J, Sorensen K (1975) A major shear zone within the Nagssugtoquidian of West Greenland. Tectonophysics 27:191–209

Barnhoorn A, Bystricky M, Burlini L, Kunze K (2004) The role of recrystallisation on the deformation behaviour of calcite rocks: Large strain torsion experiments on Carrara marble. J Struct Geol 26:885–903

Baratoux L, Schulmann K, Ulrich S, Lexa O (2005) Contrasting microstructures and deformation mechanisms in metagabbro mylonites contemporaneously deformed under different temperatures (c. 650 °C and c. 750 °C). Geological Society Special Publication 243:97–125

Behrmann JH (1987) A precautionary note on shear bands as kinematic indicators. J Struct Geol 9:659–666

Behrmann JH, Mainprice D (1987) Deformation mechanisms in a high-temperature quartz feldspar mylonite: evidence for superplastic flow in the lower continental crust. Tectonophysics 140:297–305

Bell TH, Etheridge MA (1973) Microstructure of mylonites and their descriptive terminology. Lithos 6:337–348

Benn K, Allard B (1989) Preferred mineral orientations related to magmatic flow in ophiolite layered gabbros. J Petrol 30:925–946

Berlenbach JW, Roering C (1992) Sheath-fold-like structures in pseudotachylytes. J Struct Geol 14:847–856

Berthé D, Choukroune P, Gapais D (1979b) Orientations préférentielles du quartz et orthogneissification progressive en régime cisaillant: l'exemple du cisaillement sud-armoricain. Bull Minéral 102:265–272

Berthé D, Choukroune P, Jegouzo P (1979a) Orthogneiss, mylonite and non-coaxial deformation of granites: the example of the South Armoricain shear zone. J Struct Geol 1:31–42

Bestmann M, Kunze K, Matthews A (2000) Evolution of a calcite marble shear zone complex on Thassos Island, Greece: microstructural and textural fabrics and their kinematic significance. J Struct Geol 22:1789–1807

Bestmann M, Prior DJ (2003) Intragranular dynamic recrystallisation in naturally deformed calcite marble: a case study by means of misorientation analysis. J Struct Geol 25:1597–1613

Bestmann M, Prior DJ, Veltkamp KTA (2004) Development of single-crystal sigma-shaped quartz porphyroclasts by dissolution-precipitation creep in a calcite marble shear zone. J Struct Geol 26:869–883

Bjørnerud MG, Magloughlin JF (2004) Pressure-related feedback processes in the generation of pseudotachylytes. J Struct Geol 26:2317–2323

Bjørnerud MG, Zhang H (1995) Flow mixing, object-matrix coherence, mantle growth and the development of porphyroclast tails. J Struct Geol 17:1347–1350

Blenkinsop TG (1991) Cataclasis and processes of particle size reduction. Pure Appl Geophys 136:59–86

Blenkinsop TG (2000) Deformation Microstructures and Mechanisms in Minerals and Rocks. Kluwer Academic Publishers, Dordrecht, 150 pp

Blenkinsop TG, Treloar P J (1995) Geometry, classification and kinematics of S-C fabrics. J Struct Geol 17:397–408

Blumenfeld P, Bouchez JL (1988) Shear criteria in granite and migmatite deformed in the magmatic and solid states. J Struct Geol 10:361–372

Bons PD, Barr TD, Ten Brink CE (1997) The development of delta-clasts in non-linear viscous materials: a numerical approach. Tectonophysics 270:29–41

Bos B, Peach CJ, Spiers CJ (2000) Frictional-viscous flow of simulated fault gouge caused by the combined effects of phyllosilicates and pressure solution. Tectonophysics 327:173–194

Bose S, Marques FO (2004) Controls on the geometry of tails around rigid circular inclusions: insights from analogue modelling in simple shear. J Struct Geol 26:2145–2156

Brunel M (1980) Quartz fabrics in shear zone mylonites: Evidence for a major imprint due to late strain increments. Tectonophysics 64:T33–T44

Brunel M (1986) Ductile thrusting in the Himalayas: shear sense criteria and stretching lineations. Tectonics 5:247– 265

Burg JP, Wilson CJL, Mitchell JC (1986) Dynamic recrystallization and fabric development during the simple shear deformation of ice. J Struct Geol 8:857–870

Büttner SH (2005) Deformation-controlled cation diffusion in compositionally zoned tourmaline. Mineralogical Magazine 69:471–489

Camacho A, Vernon RH, Fitz Gerald JD (1995) Large volumes of anhydrous pseudotachylyte in the Woodroffe Thrust, eastern Musgrave Ranges, Australia. J Struct Geol 17:371–383

Carreras J, Druguet E, Griera A (2005) Shear zone-related folds. J Struct Geol 27:1229–1251

Cashman S, Cashman K (2000) Cataclasis and deformation-band formation in unconsolidated marine terrace sand, Humboldt County, California. Geology 28:111–114

Ceriani S, Mancktelow NS, Pennacchioni G (2003) Analogue modelling of the influence of shape and particle/matrix interface lubrication on the rotational behaviour of rigid particles in simple shear. J Struct Geol 25: 2005–2021

Chester FM, Evans JP, Biegel RL (1993) Internal structure and weakening mechanisms of the San Andreas fault. Journal of Geophysical Research 98:771–786

Chester FM, Friedman M, Logan JM (1985) Foliated cataclasites. Tectonophysics 111:139–146

Chester FM, Logan JM (1987) Composite planar fabric of gouge from the Punchbowl Fault, California. J Struct Geol 9:621–634

Chester FM, Logan JM (1998) Shear band fabric in gouge. In: Snoke A, Tullis J, Todd VR (eds) Fault related rocks – a photographic atlas. Princeton University Press, New Jersey:68–69

Chester M, Chester JS (1998) Ultracataclasite structure and friction processes of the San Andreas fault. Tectonophysics 295:199–221

Choukroune P, Lagarde JL (1977) Plans de schistosité et déformation rotationelle: l'exemple du gneiss de Champtoceaux (Massif Armoricain). CR Acad Sci Paris 284:2331–2334

Christie JM (1958) Dynamic interpretation of the fabric of a dolomite from the Moine thrust-zone in north-west Scotland. Am J Sci 256:159–170

Cladouhos TT (1999a) Shape preferred orientations of survivor grains in fault gouge. J Struct Geol 21:419–436

Cladouhos TT (1999b) A kinematic model for deformation within brittle shear zones. J Struct Geol 21:437–448

Clarke GL (1990) Pyroxene microlites and contact metamorphism in pseudotachylyte veinlets from MacRobertson Land, East Antarctica. Aust J Earth Sci 37:1–8

Clarke GL, Norman A (1993) Generation of pseudotachylite under granulite facies conditions, and its preservation during cooling. Journal of Metamorphic Geology 11:319–335

Cobbold P (1976) Mechanical effects of anisotropy during large finite deformations. Bull Soc Geol Fr 18:1497–1510

Cobbold PR, Cosgrove JW, Summers JM (1971) Development of internal structures in deformed anisotropic rocks. Tectonophysics 12:23–53

Cobbold PR, Quinquis H (1980) Development of sheath folds in shear regimes. J Struct Geol 2:119–126

Coelho S, Passchier CW, Grasemann B (2005) Geometric description of flanking structures. J Struct Geol 27– in press

Coelho S, Passchier CW, Marques F (2006) Riedel-shear control on the development of pennant veins: Field example and analogue modelling. J Struct Geol 28:1658–1669

Conrad RE, Friedman M (1976) Microscopic feather fractures in the faulting process. Tectonophysics 33:187–198

Cox SF (1995) Faulting processes at high fluid pressures: An example of fault valve behavior from the Wattle Gully fault, Victoria , Australia: Journal of Geophysical Research 100:12841–12859

Cox SJ, Wall VJ, Etheridge MA, Potter TF (1991) Deformation and metamorphic processes in the formation of mesothermal, vein-hosted gold deposits: examples from the Lachlan fold belt in central Victoria, Australia. Ore Geology Reviews 6:391–423

Curewitz D, Karson JA (1999) Ultracataclasis, sintering, and frictional melting in pseudotachylytes from East Greenland. J Struct Geol 21:1693–1713

Davis GH, Gardulski AF, Lister GS (1987) Shear zone origin of quartzite mylonite and mylonitic pegmatite in the Coyote Mountains metamorphic core complex, Arizona. J Struct Geol 9:289–297

De Bresser JHP (1989) Calcite c-axis textures along the Gavarnie thrust zone, central Pyrenees. Geol Mijnb 68:367–376

Dell'Angelo LN, Tullis J (1989) Fabric development in experimentally sheared quartzites. Tectonophysics 169:1–21

Dennis AJ, Secor DT (1987) A model for the development

of crenulations in shear zones with applications from the Southern Appalachian Piedmont. J Struct Geol 9:809–817
Di Toro G, Pennacchioni G (2004) Superheated friction-induced melts in zoned pseudotachylytes within the Adamello tonalites (Italian Southern Alps). J Struct Geol 26:1783–1801
Doherty R (1980) Dendritic growth. In: Hargraves, RB (ed) Physics of magmatic processes. Princeton University Press, New Jersey: 576–600
Du Bernard X, Eichhubl P, Aydin A (2002) Dilatation bands: A new form of localized failure in granular media. Geophys Res Lett 29/24, doi: 10.1029/2002GL015966
Duebendorfer M, Christensen CH (1998) Plastic-to-brittle deformation of microcline during deformation and cooling of a granitic pluton. In: Snoke A, Tullis J, Todd VR (eds) Fault related rocks – a photographic atlas. Princeton University Press, New Jersey:176–179
Ebert A, Herwegh M, Berger A, Pfiffner A (2008) Grain coarsening maps for polymineralic carbonate mylonites: A calibration based on data from different Helvetic nappes (Switzerland). Tectonophysics 457:128–142
Ebert A, Herwegh M, Evans B, Pfiffner A, Austin N, Vennemann T (2007) Microfabrics in carbonate mylonites along a large-scale shear zone (Helvetic Alps). Tectonophysics 444:1–26
Eisbacher GH (1970) Deformation mechanics of mylonitic rocks and fractured granites in Cobequid Mountains, Nova Scotia. Can Bull Geol Soc Am 81:2009–2020
Ermanovics IF, Helmstaedt H, Plant AG (1972) An occurrence of Archean pseudotachylite for southeastern Manitoba. Canadian Journal of Earth Sciences 9:257–265.
Esteban JJ, Cuevas J, Vegas N, Tubía JM (2008) Deformation and kinematics in a melt-bearing shear zone from the Western Betic Cordilleras (Southern Spain) J Struct Geol 30:380–393
Evans JP (1988) Deformation mechanisms in granitic rocks at shallow crustal levels. J Struct Geol 10:437–443
Evans JP (1990) Textures, deformation mechnisms and the role of fluids in the cataclastic deformation of granite rocks. Spec Publ Geol Soc Lond 54:29–39
Evans JP (1998) Deformation of granitic rocks at shallow crustal levels. In: Snoke A, Tullis J, Todd VR (eds) Fault related rocks – a photographic atlas. Princeton University Press, New Jersey:38–39
Evans P, Chester FM (1995) Fluid-rock interaction and weakening of fault of the San Andreas system: Inferences from San Gabriel fault rock geochemistry and microstructures. Journal of Geophysical Research 100:13007–13020
Exner U, Mancktelow NS, Grasemann B (2004) Progressive development of s-type flanking folds in simple shear. J Struct Geol 26:2191–2201
Fabbri O, Lin A, Tokushige H (2000) Coeval formation of cataclasite and pseudotachylyte in a Miocene forearc granodiorite, southern Kyushu, Japan. J Struct Geol 22:1015–1025
Fernandez A, Feybesse JR, Mezure JF (1883) Theoretical and experimental study of fabric development by different shaped markers in two-dimensional simple shear. Bull Soc Geol Fr 25:319–326
Ferreira JM, Bezerra FHR, Sousa MOL, Nascimento AF, Sá JM, França GS (2008) The role of Precambrian mylonitic belts and present-day stress field in the coseismic reactivation of the Pernambuco lineament, Brazil. Tectonophysics 456:111–126
Fisher QJ, Knipe RJ (2001) The permeability of faults within siliciclastic petroleum reservoirs of the North Sea and Norwegian Continental Shelf. Mar Petrol Geol 18:1063–1081
Fliervoet TF, White SH, Drury MR (1997) Evidence for dominant grain-boundary sliding deformation in greenschist- and amphibolite-grade polymineralic ultramylonites from the Redbank Deformed Zone, Central Australia. J Struct Geol 19:1495–1520
Fredrich J, Evans B (1992) Strength recovery along simulated faults by solution transfer processes. In: Tillerson JR, Wawersik WR (eds), Rock mechanics. Balkema, Rotterdam: 121–130
Friedman M, Logan JM (1970) Microscopic Feather Fractures. Bull Geol Soc Am 81:3417-3420
Furusho M, Kanagawa K (1999) Transformation-induced strain localization in a lherzolite mylonite from the Hidaka metamorphic belt of central Hokkaido, Japan. Tectonophysics 313:411–432
Fusseis F, Handy MR (2008) Micromechanisms of shear zone propagation at the brittle–viscous transition. J Struct Geol 30:1242–1253
Fusseis F, Handy MR, Schrank C (2006) Networking of shear zones at the brittle-to-viscous transition (Cap de Creus, NE Spain). J Struct Geol 28:1228–1243
Gapais D, White SH (1982) Ductile shear bands in a naturally deformed quartzite. Text Microstruct 5:1–17
Gleason GC, DeSisto S (2008) A natural example of crystal-plastic deformation enhancing the incorporation of water into quartz. Tectonophysics 446:16–30
Goscombe B, Passchier CW (2003) Asymmetric boudins as shear sense indicators – an assessment from field data. J Struct Geol 25:575–589
Grasemann B, Stüwe K (2001) The development of flanking folds during simple shear and their use as kinematic indicators. J Struct Geol 23:715–724
Grasemann B, Stüwe K, Vannay JC (2003) Sense and non-sense of shear in flanking structures. J Struct Geol 25:19–34
Griggs DT, Turner FJ, Heard HC (1960) Deformation of rocks at 500°–800 °C. In: Griggs DT, Handin J (eds) Rock deformation. Mem Geol Soc Am 79:39–104
Grocott J (1977) The relationship between Precambrian shear belts and modern fault systems. J Geol Soc Lond 133:257–262
Grocott J (1981) Fracture geometry of pseudotachylite generation zones-a study of shear fractures formed during seismic events. J Struct Geol 3:169–179
Guermani A, Pennacchioni G (1998) Brittle precursors of plastic deformation in a granite: an example from the Mont Blanc massif (Helvetic, western Alps). J Struct Geol 20:135–148
Hafner M, Passchier CW (2000) Development of S-C′ type cleavage in paraffin waxusing a circular shear rig. Journal of the Virtual Explorer 2:16–17
Halfpenny A, Prior DJ, Wheeler J (2006) Analysis of dynamic recrystallization and nucleation in a quartzite mylonite. Tectonophysics 427:3–14
Handy MR, Wissing SB, Streit LE (1999) Frictional-viscous flow in mylonite with varied bimineralic composition and its effect on lithospheric strength. Tectonophysics 303:175–191

Hanmer S (1984a) Strain-insensitive foliations in polymineralic rocks. Can J Earth Sci 21:1410–1414
Hanmer S (1984b) The potential use of planar and elliptical structures as indicators of strain regime and kinematics of tectonic flow. Geol Surv Can Pap 84:133–142
Hanmer S (1988) Great Slave Lake Shear Zone, Canadian Shield: reconstructed vertical profile of a crustal-scale fault zone. Tectonophysics 149:245–264
Hanmer S (1990) Natural rotated inclusions in non-ideal shear. Tectonophysics 176:245–255
Hanmer S (2000) Matrix mosaics, brittle deformation, and elongate porphyroclasts: granulite facies microstructures in the Striding-Athabasca mylonite zone, western Canada. J Struct Geol 22:947–967
Hanmer S, Passchier CW (1991) Shear sense indicators: a review. Geol Surv Can Pap 90:1–71
Hanmer S, Williams M, Kopf C (1995) Modest movements, spectacular fabrics in an intracontinental deep-crustal strike-slip fault: Striding- Athabasca mylonite zone, NW Canadian Shield. J Struct Geol 17:493–507
Harris LB, Cobbold PR (1985) Development of conjugate shear bands during bulk simple shearing. J Struct Geol 7:37–44
Hartwig G (1925) Praktisch-geologische Beschreibung des Kalisalzbergwerkes "Rössing-Barnten" bei Hildesheim. Jber Niedersächs Geol Ver 17:1–74
Herwegh M, de Bresser JHP, ter Heege JH (2005) Combining natural microstructures with composite flow laws: An improved approach for the extrapolation of lab data to nature. J Struct Geol 27:503–521
Herwegh M, Handy M, Heilbronner R (2000) Evolution of mylonitic microfabrics. In: Jessel MW, Urai JL (eds) Stress, Strain and Structure, A volume in honour of WD Means. Journal of the Virtual Explorer 2
Herwegh M, Handy M, Panazzo Heilbronner R (1997) Temperature and strain rate dependent microfabric evolution in monomineralic mylonite: evidence from in-sutu deformation of norcamphor. Tectonophysics 280:83–106
Herwegh M, Handy MR (1996) The evolution of high-temperature mylonitic microfabrics: evidence from simple shearing of a quartz analogue (norcamphor). J Struct Geol 18:689–710
Herwegh M, Handy MR (1998) The origin of shape preferred orientations in mylonite: inferences from in-situ experiments on polycrystalline norcamphor. J Struct Geol 20:681–694
Herwegh M, Handy MR, Heilbronner R (1997) Temperature- and strain-rate-dependent microfabric evolution in monomineralic mylonite: evidence from in situ deformation of norcamphor. Tectonophysics 280:83–106
Herwegh M, Xiao X, Evans B (2003) The effect of dissolved magnesium on diffusion creep in calcite. Earth PlanetScieLett 212:457–470
Hippertt J F M (1993) 'V'-pull-apart microstructures: a new shear sense indicator. J Struct Geol 15: 1393–1404
Hippertt J, Rocha A, Lana C, Egydio-Silva M, Takeshita T (2001) Quartz plastic segregation and ribbon development in high-grade striped gneisses. J Struct Geol 23: 67–80
Hippertt JF, Hongn FD (1998) Deformation mechanisms in the mylonite/ultramylonite transition. J Struct Geol 20:1435–1448
Hirose T, Hayman NW (2008) Structure, permeability, and strength of a fault zone in the footwall of an oceanic core complex, the Central Dome of the Atlantis Massif, Mid-Atlantic Ridge, 30°N. J Struct Geol 30:1060–1071
Hisada E (2004) Clast-size analysis of impact-generated pseudotachylite from Vredefort Dome, South Africa. J Struct Geol 26:1419–1424
Hobbs BE, Means WD, Williams PF (1976) An outline of structural geology. Wiley, New York
Hooper RJ, Hatcher RD (1988) Mylonites from the Towaliga fault zone, central Georgia: products of heterogeneous non-coaxial deformation. Tectonophysics 152:1–17
Hutton DHW (1982) A tectonic model for the emplacement of the Main Donegal granite, NW Ireland. J Geol Soc Lond 139:615–631
Iacopini D, Passchier CW, Koehn D, Carosi R (2007) Fabric attractors in general triclinic flow systems and their application to high strain shear zones: A dynamical system approach. J Struct Geol 29:298–317
Ildefonse B, and Mancktelow NS (1993) Deformation around rigid particles: The influence of slip at the particle/matrix interface. Tectonophysics 221:345–359
Ishii K, Kanagawa K, Shigematsu N, Okudaira T (2007) High ductility of K-feldspar and development of granitic banded ultramylonite in the Ryoke metamorphic belt, SW Japan. J Struct Geol 29:1083–1098
Jefferies SP, Holdsworth RE, Wibberly CAJ, Shimamoto T, Spiers CJ, Niemeijer AR, Lloyd GE (2006) The nature and importance of phyllonite development in crustal-scale fault cores: an example from the Median Tectonic Line, Japan. J Struct Geol 28:220–235
Jezek J, Melky R, Schulmann K, Venera Z (1994) The behaviour of rigid triaxial ellipsoidal particles in viscous flows-modeling of fabric evolution in a multiparticle system. Tectonophysics 229:165–180
Ji S, Jiang Z, Rybacki E, Wirth R, Prior D, Xia B (2004) Strain softening and microstructural evolution of anorthite aggregates and quartz-anorthite layered composites deformed in torsion. Earth and Planetary Science Letters 222:377–390
Jiang D (2007) Sustainable transpression: An examination of strain and kinematics in deforming zones with migrating boundaries. J Struct Geol 29:1984–2005
Kanaori Y, Kawakami S, Yairi K (1991) Microstructure of deformed biotite defining foliation in cataclasite zones in granite, central Japan. J Struct Geol 13:777–786
Kanagawa K, Shimano H, Hiroi Y (2008) Mylonitic deformation of gabbro in the lower crust: A case study from the Pankenushi gabbro in the Hidaka metamorphic belt of central Hokkaido, Japan. J Struct Geol 30:1150–1166
Kano K, Sato H (1988) Foliated fault gouges: examples from the shear zones of the Sakai-Toge and Narai faults, central Japan. J Geol Soc Jpn 94:453–456
Kellermann Slotemaker A, de Bresser JHP (2006) On the role of grain topology in dynamic grain growth – 2D microstructural modeling. Tectonophysics 427:73–93
Kenkmann T (2000) Processes controlling the shrinkage of porphyroclasts in gabbroic shear zones. J Struct Geol 22:471–487
Killick AM, Thwaites AM, Germs GJB, Schoch AE (1988) Pseudotachylite associated with a bedding-parallel fault zone between the Witwatersrand and Ventersdorp Supergroups, South Africa. Geologische Rundschau 77:329–344
Knipe RJ, Law RD (1987) The influence of crystallographic orientation and grain boundary migration on

microstructural and textural development in a S-C mylonite. Tectonophysics 135:155–169.
Krohe A (1990) Local variations in quartz (c)-axis orientations in non-coaxial regimes and their significance for the mechanics of S-C fabrics. J Struct Geol 12:995–1004
Kronenberg AK (1994) Hydrogen speciation and chemical weakening of quartz. In: Heaney PJ, Prewitt CT, Gibbs GV (eds) Silica: Physical behavior, geochemistry, and materials applications. Mineral. Soc. Am., Rev. Mineral. 29:123–176
Kronenberg AK, Kirby SH, Pikston JC (1990) Basal slip and mechanical anisotropy of biotite. J Geophys Res 95:19257–19278
Kruse R, Stünitz H (1999) Deformation mechanisms and phase distribution in mafic high-temperature mylonites from the Jotun Nappe, Southern Norway. Tectonophysics 303:223–249
Kuiper YD, Jiang D, Lin S (2007) Relationship between non-cylindrical fold geometry and the shear direction in monoclinic and triclinic shear zones. J Struct Geol 29:1022–1033
Kurz GA, Northrup CJ (2008) Structural analysis of mylonitic rocks in the Cougar Creek Complex, Oregon–Idaho using the porphyroclast hyperbolic distribution method, and potential use of SC'-type extensional shear bands as quantitative vorticity indicators. J Struct Geol 30: 1005–1012
Lacassin R, Mattauer M (1985) Kilometre scale sheath fold at Mattmark and implications for transport direction in the Alps. Nature 315:739–742
Lafrance B, Vernon RH (1998) Coupled mass transfer and microfracturing in gabbroic mylonites. In: Snoke A, Tullis J, Todd VR (eds) Fault related rocks – a photographic atlas. Princeton University Press, New Jersey:204–207
Lapworth C (1885) The highland controversy in British geology: its causes, course and consequences. Nature 32:558–559
Laurent P (1987) Shear-sense determination on striated faults from e twin lamellae in calcite. J Struct Geol 9:591–596
Law RD (1998) Oblique grain-shape fabrics in a mylonitic quartz vein. In: Snoke A, Tullis J, Todd VR (eds) Fault related rocks – a photographic atlas. Princeton University Press, New Jersey:264–265
Law RD, Knipe RJ, Dayan H (1984) Strain path partitioning within thrust sheets: microstructural and petrofabric evidence from the Moine Thrust zone at Loch Eriboll, northwest Scotland. J Struct Geol 6:477–497
Legros F, Cantagrel JM, Devouard B (2000) Pseudotachylyte at the base of the Arequipa volcanic landslide deposit: implications for emplacement mechanisms. Geology 108:601–611
Lenze A, Stöckhert B (2007) Microfabrics of UHP metamorphic granites in the Dora Maira Massif, western Alps – No evidence of deformation at great depth. J metam Geol 25:461–475
Lin A (1994) Glassy pseudotachylyte veins from the Fuyun fault zone, northwest China. J Struct Geol 16:71–84
Lin A (1996) Injection veins of crushing-originated pseudotachylyte and fault gouge formed during seismic faulting. Engineering Geology 43:213–224
Lin A (1997) Ductile deformation of biotite in foliated cataclasites, Iida-Matsukawa fault, central Japan. Journal of Asian Earth Sciences 15:407–411
Lin A (1998) Glassy and microlitic pseudotachylytes In: Snoke A, Tullis J, Todd VR (eds) Fault related rocks – a photographic atlas. Princeton University Press, New Jersey: 112–121
Lin A (1999) S-C cataclasite in granitic rock. Tectonophysics 304: 257–273
Lin A (2001) S-C fabrics developed in cataclastic rocks from the Nojima fault zone, Japan and their implications for tectonic history. J Struct Geol 23:1167–1178
Lin A, Miyata T, Wan T (1998) Tectonic characteristics of the central segment of the Tancheng-Lujiang fault zone, Shandong Peninsula, eastern China. Tectonophysics 293:85–104
Lin A, Shimamoto T (1998) Selective melting processes as inferred from experimentally generated pseudotachylytes. Journal of Asian Earth Sciences 16:533–545
Lin S, Jiang D, Williams PF (2007) Importance of differentiating ductile slickenside striations from stretching lineations and variation of shear direction across a high-strain zone. J Struct Geol 29:850–862
Lister GS, Snoke AW (1984) S-C Mylonites. J Struct Geol 6:617–638
Liu R, Dai L, Wang T, Cui W, Yang X (2007) Deformation partitioning in ductile shear zones – An example from the Dandong ductile shear zone. Scientia Geologica Sinica 42:223–233
Lofgren GE (1971a) Spherulitic structures in glassy and crystalline rocks. Journal of Geophysical Research 76:5635–5648
Lofgren GE (1971b) Experimentally produced devitrification textures in natural rhyolite glass. Geological Society of America Bulletin 82:111–124
Lofgren GF (1974) an experimental study of plagioclase crystal morphology: isothermal crystallization. American Journal of Science 247:243–273
Logan JM, Dengo CA, Higgs NG, Wang ZZ (1992) Fabrics of experimental fault zones: Their development and relationship to mechanical behavior. In: Evans B, Wong TF (eds) Fault mechanics and transport properties of rocks. New York Academic Press 33–68
Logan JM, Friedman M, Higgs NG, Dengo C, Shimamoto T (1979) Experimental studies of simulated gouge and their application to studies of natural fault zones. US Geol Surv Open-file Rep 79–1239:305–343
Lotze F (1957) Steinsalz und Kalisaltze I. Gebrüder Bornträger, Berlin 465
Luan FC, Paterson MS (1992) Preparation and deformation of synthetic aggregates of quartz. J Geophys Res 97:301–320
Luneau P, Cruden AR, (1998) magmatic fabric acquisition mechanisms in a syenite: results of a combined anisotropy of magnetic susceptibility and image analysis study. J Geophys Res 103:5067–5089
Macaudiere J, Brown WL, Ohnenstetter D (1985) Microcrystalline textures resulting from rapid crystallization in a pseudotachylite melt in a meta-anorthosite. Contrib Mineral Petrol 89:39–51
Maddock RH (1986) Partial melting of lithic porphyroclasts in fault-generated pseudotachylites. Neues Jahrb Miner Abh 155:1–14
Maddock RH (1992) Effects of lithology, cataclasis and melting on the composition of fault-generated pseudotachylytes in Lewisian gneiss, Scotland. Tectonophysics 204:261–278

Maddock RH (1998) Pseudotachylyte formed by frictional fusion. In: Snoke A, Tullis J, Todd VR (eds) Fault related rocks – a photographic atlas. Princeton University Press, New Jersey: 80–87

Maddock RH, Grocott J, van Nes M (1987) Vesicles, amygdales and similar structures in fault-generated pseudotachylites. Lithos 20:419–432

Magloughlin JF (1989) The nature and significance of pseudotachylite from the Nason terrane, North Cascade Mountains, Washington. J Struct Geol 11:907–917

Magloughlin JF (1992) Microstructural and chemical changes associated with cataclasis and friction melting at shallow crustal levels: The cataclasite-pseudotachylyte connection. In: Magloughlin JF and Spray JG (eds) Frictional Melting Processes and Products in Geological Materials. Tectonophysics, 204: 243–260.

Magloughlin JF (1998a) Amygdules within pseudotachylyte. In: Snoke A, Tullis J, Todd VR (eds) Fault related rocks – a photographic atlas. Princeton University Press, New Jersey:92–93

Magloughlin JF (1998b) Amygdules and microbreccia collapse structure in a pseudotachylyte. In: Snoke A, Tullis J, Todd VR (eds) Fault related rocks – a photographic atlas. Princeton University Press, New Jersey:94–95

Magloughlin JF, Spray JG (1992) Frictional melting processes and products in geological materials: introduction and discussion. Tectonophysics 204:197–204

Main I, Mair K, Kwon O, Elphick S, Ngwenya B (2001) Experimental constraints on the mechanical and hydraulic properties of deformation bands in porous sandstones; a review. In: Holdsworth RE, Strachan RA, Magloughlin JF, Knipe RJ (eds) The nature and tectonic significance of fault zone weakening. Geol Soc Special Pub 186, pp 43–63

Mair K, Elphick S, Main I (2002) Influence of confining pressure on the mechanical and structural evolution of laboratory deformation bands. Geophys Res Lett 29 doi: 10.1029/2001GL013964

Mair K, Main I, Elphick S (2000) Sequential growth of deformation bands in the laboratory. J Struct Geol 22:25–42

Malavieille J, Cobb F (1986) Cinématique des déformations ductiles dans trois massifs métamorphiques de l'Ouest des Etats-Unis: Albion (Idaho), Raft River et Grouse Creek (Utah). Bull Soc Geol France 2:885–898

Mancktelow NS, Arbaret L, Pennacchioni G (2002) Experimental observations on the effect of interface slip on rotation and stabilisation of rigid particles in simple shear and a comparison with natural mylonites. J Struct Geol 24:567–585

Marques FO, Coelho S (2001) Rotation of rigid elliptical cylinders in viscous simple shear flow: Analogue experiments. J Struct Geol 23:609–617

Marques FO, Guerreiro SM, Fernandes AR (2008) Sheath fold development with viscosity contrast: Analogue experiments in bulk simple shear. J Struct Geol 30:1348–1358

Marques FO, Schmid DW, Andersen TB (2007) Applications of inclusion behaviour models to a major shear zone system: The Nordfjord-Sogn Detachment Zone in western Norway. J Struct Geol 29:1622–1631

Martini JEJ (1992) The metamorphic history of the Vredefort dome at approximately 2 Ga as revealed by coesite-stishovite-bearing pseudotachylites. Journal of Metamorphic Geology 10:517–527

Masch L, Wenk JT, Preuss E (1985) Electron microscopy study of hyalomylonites – evidence for frictional melting in landslides. Tectonophysics 115:131–160

Masuda T, Michibayashi K, Otha H,(1995) Shape preferred orientation of rigid particles in a viscous matrix: re-evaluation to determine kinematic parameters of ductile deformation. J Struct Geol 17:115–129

McCaig A (1998) Fluid flow mechanisms in mylonites: Evidence from compositional zoning patterns In: Snoke A, Tullis J, Todd VR (eds) Fault related rocks – a photographic atlas. Princeton University Press, New Jersey:208–209

McCaig AM (1987) Deformation and fluid-rock interaction in metasomatic dilatant shear bands. Tectonophysics 135:121–132

McLelland J (1984) The origin of ribbon lineation within the southern Adirondacks, USA. J Struct Geol 6:147–157

Means WD (1981) The concept of steady – state foliation. Tectonophysics 78:179–199

Means WD (1987) A newly recognized type of slickenside striation. J Struct Geol 9:585–590

Mehl L, Hirth G (2008) Plagioclase preferred orientation in layered mylonites: Evaluation of flow laws for the lower crust

Journal of Geophysical Research B: Solid Earth 113 art. no. B05202

Menegon L, Pennacchioni G, Heilbronner R, Pittarello L (2008) Evolution of quartz microstructure and c-axis crystallographic preferred orientation within ductilely deformed granitoids (Arolla unit, Western Alps) J Struct Geol 30:1332–1347

Menegon L, Pennacchioni G, Spiess R (2008) Dissolution-precipitation creep of K-feldspar in mid-crustal granite mylonites. J Struct Geol 30:565–579

Menéndez B, Zhu W, Wong TF (1996) Micromechanics of brittle faulting and cataclastic flow in Berea sandstone. J Struct Geol 18:1–16

Michibayashi K, Murakami M (2007) Development of a shear band cleavage as a result of strain partitioning. J Struct Geol 29:1070–1082

Mitra G (1978) Ductile deformation zones and mylonites: the mechanical processes involved in the deformation of crystalline basement rocks. Am J Sci 278:1057–1084

Mollema PN, Antonellini MA (1996) Compaction bands: a structural analog for anti-mode I cracks in aeolian sandstone. Tectonophysics 267:209–228

Mort K, Woodcock NH (2008) Quantifying fault breccia geometry: Dent Fault, NW England. J Struct Geol 30:701–709

Nakashima S, Matayoshi H, Yuko T, Michibayashi K, Masuda T, Kuroki N, Yamagishi H, Ito Y, Nakamura A (1995) Infrared microspectroscopy analysis of water distribution in deformed and metamorphosed rocks. Tectonophysics 245:263–276

Newman J, Lamb WM, Drury MR, Vissers RLM (1999) Deformation processes in a peridotite shear zone: reaction-softening by an H_2O-deficient, continuous net transfer reaction. Tectonophysics 303:193–222

Norrell GT, Teixell A, Harper GD (1989) Microstructure of serpentinite mylonites from the Josephine ophiolite and serpentinitization in retrogressive shear zones, California. Bull Geol Soc Am 101:673–682

Oesterling N, Heilbronner R, Stünitz H, Barnhoorn A, Molli G (2007) Strain dependent variation of

microstructure and texture in naturally deformed Carrara marble. J Struct Geol 29:681–696
O'Hara K (1992) Major- and trace-element constraints on the petrogenesis of a fault-related pseudotachylyte, western Blue Ridge province, North Carolina. In: Magloughlin JF, Spray JG (eds) Frictional Melting Processes and Products in Geological Materials. Tectonophysics 204:279–288
O'Hara K (2001) A pseudotachylyte geothermometer. J Struct Geol 23:1345–1357
O'Brien DK, Wenk HR, Ratschbacher L, You Z (1987) Preferred orientation of phyllosilicates in phyllonites and ultramylonites. J Struct Geol 9:719–730
Ogilvie SR, Glover PWJ (2001) The petrophysical properties of deformation bands in relation to their microstructure. Earth and Planetary Scientific Letters 193:129–142
Olesen NØ (2008) The microfabrics of a porphyroclast-rich quartzitic mylonite, Mjølfjell, Jotun Nappe Complex, Norway. Norsk Geologisk Tidsskrift 88:89–101
Oliver DH, Goodge JW (1996) Leucoxene fish as a micro-kinematic indicator. J Struct Geol 18:1493–1495
Passchier CW (1982a) Mylonitic deformation in the Saint-Barthélemy Massif, French Pyrenees, with emphasis on the genetic relationship between ultramylonite and pseudotachylyte. GUA Pap Geol Ser 1 16:1–173
Passchier CW (1982b) Pseudotachylyte and the development of ultramylonite bands in the Saint-Barthélemy Massif, French Pyrenees. J Struct Geol 4:69–79
Passchier CW (1984) The generation of ductile and brittle shear bands in a low-angle mylonite zone. J Struct Geol 6:273–281
Passchier CW (1985) Water deficient mylonite zones – an example from the Pyrenees. Lithos 18: 115–127
Passchier CW (1986a) Mylonites in the continental crust and their role as seismic reflectors. Geol Mijnb 65:167–176
Passchier CW (1986b) Flow in natural shear zones: the consequences of spinning flow regimes. Earth Plan et Sci Lett 77:70–80
Passchier CW (1987a) Efficient use of the velocity gradients tensor in flow modelling. Tectonophysics 136:159–163
Passchier CW (1987b) Stable positions of rigid objects in non-coaxial flow: a study in vorticity analysis. J Struct Geol 9:679–690
Passchier CW (1988a) Analysis of deformation paths in shear zones. Geol Rdsch 77:309–318
Passchier CW (1990a) Reconstruction of deformation and flow parameters from deformed vein sets. Tectonophysics 180:185–199
Passchier CW (1991b) Geometric constraints on the development of shear bands in rocks. Geol Mijnb 70:203–211
Passchier CW (1994) Mixing in flow perturbations: a model for development of mantled porphyroclasts in mylonites. J Struct Geol 16:733–736
Passchier CW (2001) Flanking structures. J Struct Geol 23:951–962
Passchier CW, Druguet E (2002) Numerical modelling of asymmetric boudinage. J Struct Geol 24:1789–1803
Passchier CW, Hoek JD, Bekendam RF, de Boorder H (1990a) Ductile reactivation of Proterozoic brittle fault rocks: an example from the Vestfold Hills. East Antarctica Prec Res 47:3–16
Passchier CW, Simpson C (1986) Porphyroclast systems as kinematic indicators. J Struct Geol 8:831–844
Passchier CW, Sokoutis D (1993) Experimental modelling of mantled porphyroclasts. J Struct Geol 15:895–910
Passchier CW, ten Brink CE, Bons PD, Sokoutis D (1993) Delta-objects as a gauge for stress sensitivity of strain rate in mylonites. Earth Plan et Sci Lett 120:239–245
Pennacchioni G, Di Toro G, Mancktelow NS (2001) Strain-insensitive preferred orientation of porphyroclasts in Mont Mary mylonites. J Struct Geol 23:1281–1298
Pennacchioni G, Fasolo L, Cecchi MM, Salasnich L (2000) Finite-element modelling of simple shear flow in Newtonian and non-Newtonian fluids around a circular rigid particle. J Struct Geol 22:683–692
Petit JP (1987) Criteria for the sense of movement on fault surfaces in brittle rocks. J Struct Geol 9:597–608
Philpotts AR (1964) Origin of pseudotachylytes. Am J Sci 262:1008–1035
Piazolo S, Bons PD, Passchier CW (2002) The influence of matrix rheology and vorticity on fabric development of populations of rigid objects during plane strain deformation. Tectonophysics 351:315–329
Piazolo S, Passchier CW 2002. Controls on lineation development in low to medium-grade shear zones: a study from the Cap de Creus peninsula, NE Spain. Journal of Structural Geology 24, 25–44
Platt JP (1984) Secondary cleavages in ductile shear zones. J Struct Geol 6:439–442
Platt JP, Vissers RLM (1980) Extensional structures in anisotropic rocks. J Struct Geol 2:397–410
Post A, Tullis J (1998) The rate of water penetration in experimentally deformed quartzite: Implications for hydrolytic weakening. Tectonophysics 295:117–137
Post AD, Tullis J, Yund RA (1996) Effects of chemical environment on dislocation creep of quartzite. J Geophys Res 101:22143–22155
Raimbourg H, Toyoshima T, Harima Y, Kimura G (2008) Grain-size reduction mechanisms and rheological consequences in high-temperature gabbro mylonites of Hidaka, Japan. Earth and Planetary Science Letters 267:637–653.
Ramsay JG (1980) Shear zone geometry: a review. J Struct Geol 2:83–101
Ramsay JG, Graham RH (1970) Strain variation in shear belts. Can J Earth Sci 7:786–813
Ray SK (1999) Transformation of cataclastically deformed rocks to pseudotachylyte by pervasion of frictional melt: Inferences from clast-size analysis. Tectonophysics 301:283–304
Ray SK (2004) Melt-clast interaction and power-law size distribution of clasts in pseudotachylytes. J Struct Geol 26:1831–1843
Ree JH (1991) An experimental steady-state foliation. J Struct Geol 13:1001–1011
Reimold WU (1995) Impact cratering – a review, with special reference to the economic importance of impact structures and the southern African impact crater record. Earth, Moon and Planets 70:21–45
Riedel W (1929) Zur Mechanik geologischer Brucherscheinungen. Zentralblatt für Mineralogie, Geologie und Paläontologie 1929B:354–368.
Romeo I, Capote R, Lunar R (2007) Crystallographic preferred orientations and microstructure of a Variscan marble mylonite in the Ossa-Morena Zone (SW Iberia). J Struct Geol 29:1353–1368
Roper PJ (1972) Structural significance of „button" or „fish scale" texture in the phyllonitic schist of the

Brevard zone. Geological Society of America Bulletin 83:853–860
Rutter EH (1976) The kinetics of rock deformation by pressure solution. Phil Trans R Soc Lond A283:203–219
Rutter EH, Maddock RH, Hall SW, White SH (1986) Comparative microstructures of natural and experimentally produced clay-bearing fault gouges. Pure Appl Geophys 124:3–30
Saltzer SD, Hodges KV (1988) The Middle Mountain shear zone, southern Idaho: kinematic analysis of an early Tertiary high-temperature detachment. Bull Geol Soc Am 100:96–103
Sawaguchi T, Ishii K (2003) Three-dimensional numerical modeling of lattice- and shape-preferred orientation of orthopyroxene porphyroclasts in peridotites. J Struct Geol 25:1425–1444
Schenk O, Urai JL, Evans B (2005) The effect of water on recrystallization behavior and grain boundary morphology in calcite-observations of natural marble mylonites. J Struct Geol 27:1856–1872
Schléder Z, Urai JL (2007) Deformation and recrystallization mechanisms in mylonitic shear zones in naturally deformed extrusive Eocene-Oligocene rocksalt from Eyvanekey plateau and Garmsar hills (central Iran). J Struct Geol 29:241–255
Schmid DW, Podladchikov YY (2004) Are isolated stable rigid clasts in shear zones equivalent to voids? Tectonophysics 384:233–242
Schmid SM (1994) Textures of geological materials: computer model predictions versus empirical interpretations based on rock deformation experiments and field studies. In: Bunge H J, Siegesmund S, Skrotzki W, Weber K (eds) Textures of geological materials. DGM Informationsges, Oberursel, pp 279–301
Schmid SM, Boland JN, Paterson MS (1977) Superplastic flow in fine-grained limestone. Tectonophysics 43:257–291
Schmid SM, Handy MR (1991) Towards a genetic classification of fault rocks: geological usage and tectonophysical implications. In: Müller DW, McKenzie JA, Weissert H (eds) Controversies in Modern Geology, Evolution of Geological Theories in Sedimentology, Earth History and Tectonics. Academic Press, London, pp 339–361
Schmid SM, Zingg A, Handy M (1987) The kinematics of movements along the Insubric Line and the emplacement of the Ivrea Zone. Tectonophysics 135:47–66
Scholz CH (1988) The brittle-plastic transition and the depth of seismic faulting. Geol Rdsch 77:319–328
Scholz C H (1990) The mechanics of earthquakes and faulting. Cambridge University Press, New York, 439 pp
Scholz CH (2002) The mechanics of earthquakes and faulting. Cambridge University press. 471pp
Sengupta S, Ghosh SK (2007) Origin of striping lineation and transposition of linear structures in shear zones. J Struct Geol 29:273–287
Shand SJ (1916) The pseudotachylyte of Parijs (Orange Free State) and its relation to „trap-shotten gneiss" and „flinty crush-rock": The Quarterly Journal of the Geological Society of London 72:198–221
Shaw CA, Allen J (2007) Field rheology and structural evolution of the Homestake shear zone, Colorado. Rocky Mountain Geology 42:31–56
Shelley D (1993) Igneous and metamorphic rocks under the microscope. Chapman and Hall, London
Shelley D (1995) Asymetric shape preferred orientations as shear-sense indicators. J Struct Geol 17:509:517
Shimamoto T (1989) The origin of S-C mylonites and a new fault-zone model. J Struct Geol 11:51–64
Shimamoto T, Nagahama H (1992) An argument against the crush origin of pseudotachylytes based on the analysis of clast-size distribution. J Struct Geol 14:999–1006
Shimizu I (2008) Theories and applicability of grain size piezometers: The role of dynamic recrystallization mechanisms. J Struct Geol 30:899–917
Sibson RH (1975) Generation of pseudotachylyte by ancient seismic faulting. Geophys J R Astr Soc 43:775–794
Sibson RH (1977) Fault rocks and fault mechanisms. J Geol Soc Lond 133:191–213
Sibson RH (1980) Transient discontinuities in ductile shear zones. J Struct Geol 2:165–171
Sibson RH (1990) Conditions for fault-valve behaviour. In: Knipe RJ, Rutter EH (eds) Deformation mechanisms, rheology and tectonics. Geol Soc Special Pub 54, pp 15–28
Sidman D, Ferré EC, Teyssier C, Jackson M (2005) Magnetic fabric and microstructure of a mylonite: Example from the Bitterroot shear zone, western Montana. Geological Society Special Publication 245:143–163
Simpson C (1985) Deformation of granitic rocks across the brittle-ductile transition. J Struct Geol 7:503–511
Simpson C, Schmid SM (1983) An evaluation of criteria to determine the sense of movement in sheared rocks. Bull Geol Soc Am 94:1281–1288
Simpson C, Wintsch RP (1989) Evidence for deformation-induced K-feldspar replacement by myrmekite. J Metam Geol 7:261–275
Smith SAF, Strachan RA, Holdsworth RE (2007) Microstructural evolution within a partitioned midcrustal transpression zone, northeast Greenland Caledonides. Tectonics 26 art. no. TC4003 .
Snoke A, Tullis J (1998) An overview of fault rocks. In: Snoke A, Tullis J, Todd VR (eds) Fault related rocks – a photographic atlas. Princeton University Press, New Jersey. pp 3–18
Snoke A, Tullis J, Todd VR (1998) Fault related rocks – a photographic atlas. Princeton University Press, New Jersey. 617 pp
Spray JG (1987) Artificial generation of pseudotachylite using friction welding apparatus: simulation of melting on a fault plane. J Struct Geol 9:49–60
Spray JG (1988) Generation and crystallization of an amphibolite shear melt: an investigation using radial friction welding apparatus. Contrib Mineral Petrol 99: 464–475
Spray JG (1992) A physical basis for the frictional melting of some rock-forming minerals. Tectonophysics 204:205–221
Spray JG (1995) Pseudotachylyte controversy: fact or fiction? Gelogy 23:1119–1122
Spray JG (1997) Superfaults. Geology 25:579–582
Spray JG, Kelley SP, Reimold WU (1995) Laser probe 40Ar/39Ar dating of coesite- and stishovite-bearing pseudotachylytes and the age of the Vredefort impact event. Meteoritics 30:335–343
Spray JG, Thompson LM (1995) Friction melt distribution in a terrestrial multi-ring impact basin. Nature 373:1119–1122
Spry A (1969) Metamorphic textures. Pergamon Press, Oxford

Song WJ, Ree JH (2007) Effect of mica on the grain size of dynamically recrystallized quartz in a quartz–muscovite mylonite. J Struct Geol 29:1872–1881
Stel H (1981) Crystal growth in cataclasites: diagnostic microstructures and implications. Tectonophysics 78:585–600
Stipp, M, Kunze K (2008) Dynamic recrystallization near the brittle-plastic transition in naturally and experimentally deformed quartz aggregates. Tectonophysics 448:77–97
Strating EHH, Vissers RLM (1994) Structures in natural serpentinite gouges. J Struct Geol 16:1205–1215
Strehlau J (1986) A discussion of the depth extent of rupture in large continental earthquakes. Earthquake Source Mechanics. Geophys Monogr 37:131–146
Sullivan WA (2008) Significance of transport-parallel strain variations in part of the Raft River shear zone, Raft River Mountains, Utah, USA. J Struct Geol 30:138–158
Sullivan WA, Law RD (2007) Deformation path partitioning within the transpressional White Mountain shear zone, California and Nevada. J Struct Geol 29:583–599
Swanson MT (1992) Fault structure, wear mechanisms and rupture processes in pseudotachylyte generation. Tectonophysics 204:223–242
Swanson MT (1998) Pseudotachylyte-bearing strike-slip faults in mylonitic rocks In: Snoke A, Tullis J, Todd VR (eds) Fault related rocks – a photographic atlas. Princeton University Press, New Jersey:104–107
Takagi H (1998) Foliated fault gouges. In: Snoke A, Tullis J, Todd VR (eds) Fault related rocks – a photographic atlas. Princeton University Press, New Jersey: 58–61
Takagi H, Goto K, Shigematsu N (2000) Ultramylonite bands derived from cataclasite and pseudotachylyte in granites, northeast Japan. J Struct Geol 22:1325–1339
Talbot CJ (1970) The minimum strain ellipsoid using deformed quartz veins. Tectonophysics 9:47–76
Tanaka H (1992) Cataclastic lineations. J Struct Geol 14:1239–1252
Ten Grotenhuis SM, Passchier CW, Bons PD (2002) The influence of strain localisation on the rotation behaviour of rigid objects in experimental shear zones. J Struct Geol 24:485–501
Ten Grotenhuis SM, Trouw RAJ, Passchier CW (2003) Evolution of mica fish in mylonitic rocks. Tectonophysics 372:1–21
Teufel LW (1981) Localization of deformation along faults; implications to fault zone permeability. Geol Soc Am 13, p 565
Thompson LM, Spray JG (1994) Pseudotachylytic rock distribution and genesis within the Sudbury impact structure. In: Dressler BO, Grieve RAF, Sharpton VL (eds) Large meteorite impacts and planetary evolution. Boulder, Colorado, Geological Society of America, Special Paper 293:275–287
Tikoff B, Teyssier C (1994) Strain and fabric based on porphyroclast interaction. J Struct Geol 16:477–491
Toy VG, Prior DJ, Norris RJ (2008) Quartz fabrics in the Alpine Fault mylonites: Influence of pre-existing preferred orientations on fabric development during progressive uplift. J Struct Geol 30:602–621
Toyoshima T (1998) Gabbro mylonite developed along a crustal-scale decollement. In: Snoke A, Tullis J, Todd VR (eds) Fault related rocks – a photographic atlas. Princeton University Press, New Jersey:426–427
Toyoshima T, (1990) Pseudotachylyte from the Main Zone of the Hidaka metamorphic belt: Hokkaido, northern Japan. Journal of Metamorphic Geology 8:507–523
Treagus SH, Lan L (2003) Simple shear of deformable square objects. J Struct Geol 25:1993–2003
Trullenque G, Kunze K, Heilbronner R, Stünitz H, Schmid SM (2006) Microfabrics of calcite ultramylonites as records of coaxial and non-coaxial deformation kinematics: Examples from the Rocher de l'Yret shear zone (Western Alps). Tectonophysics 424:69–97
Treagus SH, Lan L (2004) Deformation of square objects and boudins. J Struct Geol 26:1361–1376
Tse ST, Rice JR (1986) Crustal earthquake instability in relation to the depth variation of frictional slip properties. Journal of Geophysical Research 91:9452–9472
Tsurumi J, Hosonuma H, Kanagawa K (2003) Strain localization due to a positive feedback of deformation and myrmekite-forming reaction in granite and aplite mylonites along the Hatagawa Shear Zone of NE Japan. J Struct Geol 25:557–574
Tullis J, Dell'Angelo L, Yund RA (1990) Ductile shear zones from brittle precursors in feldspathic rocks: the role of dynamic recrystallization. In: Hobbs BE, Heard HC (eds) Mineral and rock deformation: laboratory studies. AGU, Geophys Monogr 56:67-81
Tullis JT, Snoke AW, Todd VR (1982) Significance of petrogenesis of mylonitic rocks. Geology 10:227–230
Underhill JR, Woodcock NH (1987) Faulting mechanisms in high-porosity sandstones; New Red Sandstone, Arran, Scotland. In: Jones ME, Preston MF (eds) Deformation of sediments and sedimentary rocks.Geological Society Special Publications 29:91–105.
Van der Wal D, Vissers RMD, Drury MR (1992) Oblique fabrics in porphyroclastic Alpine peridotites: a shear sense indicator for upper mantle flow. J Struct Geol 14:839–846
Vernon RH (2004) A practical guide to rock microstructures. Cambridge University Press, 594 pp
Vernon RH, Williams VA, D'Arcy WF (1983) Grain-size reduction and foliation development in a deformed granitoid batholith. Tectonophysics 92:123–145
Vernooij MGC, Kunze K, den Brok B (2006) 'Brittle' shear zones in experimentally deformed quartz single crystals. J Struct Geol 28:1292–1306
Vos IMA, Bierlein FP, Barlow MA, Betts PG (2006) Resolving the nature and geometry of major fault systems from geophysical and structural analysis: The Palmerville Fault in NE Queensland, Australia. J Struct Geol 28:2097–2108
Wallace R-C (1976) Partial fusion along the Alpine Fault Zone, New Zealand. Geol Soc Am Bull 87:1225–122
Wallis SR (1992a) Vorticity analysis in a metachert from the Sanbagawa belt , SW Japan. J Struct Geol 14:271–280
Wang Z, Cheng Q, Cao L, Xia Q, Chen Z (2007) Fractal modelling of the microstructure property of quartz mylonite during deformation process. Mathematical Geology 39:53–68
Waters-Tormey C, Tikoff B (2007) Characteristics of a kilometer-scale high strain zone in the lower continental crust: Mt. Hay block, central Australia. J Struct Geol 29:562–582
Wenk H-R, Monteiro PJM, Shomglin K (2008) Relationship between aggregate microstructure and mortar expansion. A case study of deformed granitic rocks from the

Santa Rosa mylonite zone. Journal of Materials Science 43:1278–1285
Wheeler J (1987b) The determination of true shear senses from the deflection of passive markers in shear zones. J Geol Soc Lond 144:73–77
White SH (1976) The role of dislocation processes during tectonic deformation with special reference to quartz. In: Strens RJ (ed) The physics and chemistry of minerals and rocks. Wiley, London, pp 75–91
White SH (1977) Geological significance of recovery and recrystallization processes in quartz. Tectonophysics 39:143–170
White SH (1979a) Grain and sub-grain size variations across a mylonite zone. Contrib Mineral Petrol 70:193–202
White SH (1979b) Large strain deformation: report on a tectonic studies group discussion meeting held at Imperial College, London; introduction. J Struct Geol 4:333–339
White SH, Burrows SE, Carreras J, Shaw ND, Humphreys FJ (1980) On mylonites in ductile shear zones. J Struct Geol 2:175–187
Whitmeyer SJ, Simpson C (2003) High strain-rate deformation fabrics characterize a kilometers-thick Paleozoic fault zone in the Eastern Sierras Pampeanas, central Argentina. J Struct Geol 25:909–922
Wiesmayr G, Grasemann B (2005) Sense and non-sense of shear in flanking structures with layer-parallel shortening: implications for fault-related folds. J Struct Geol 27:249–264
Williams PF, Jiang D (2005) An investigation of lower crustal deformation: Evidence for channel flow and its implications for tectonics and structural studies. J Struct Geol 27:1486–1504
Wintsch RP (1998) Strengthening of fault breccia by K-feldspar cementation. In: Snoke A, Tullis J, Todd VR (eds) Fault related rocks – a photographic atlas. Princeton University Press, New Jersey:42–43
Wintsch RP, Christoffersen R, Kronenberg AK (1995) Fluid-rock reaction weakening of fault zones. Journal of Geophysical Research 100:13021–13032
Wise DU, Dunn DE, Engelder JT, Geiser PA, Hatcher RD, Kish SA, Odom AL, Schamel S (1984) Fault-related rocks: Suggestions for terminology. Geology 12:391–394
Zechmeister MS, Ferré EC, Cosca MA, Geissman JW (2007) Slow and fast deformation in the Dora Maira Massif, Italian Alps: Pseudotachylytes and inferences on exhumation history. J Struct Geol 29:1114–1130
Zhang S, Karato S (1995) Lattice preferred orientation of olivine aggregates deformed in simple shear. Nature 375:774–777
Zhao ZY, Fang AM (2005) Plastic flow of ultrahigh pressure metamorphic rocks: Microstructure and deformation mechanisms. Acta Petrologica Sinica 21:1109–1116

MIX
Papier aus verantwortungsvollen Quellen
Paper from responsible sources
FSC® C105338

If you have any concerns about our products,
you can contact us on
ProductSafety@springernature.com

In case Publisher is established outside the EU,
the EU authorized representative is:
Springer Nature Customer Service Center GmbH
Europaplatz 3, 69115 Heidelberg, Germany

Printed by Libri Plureos GmbH
in Hamburg, Germany